SpringerBriefs in Law

SpringerBriefs present concise summaries of cutting-edge research and practical applications across a wide spectrum of fields. Featuring compact volumes of 50 to 125 pages, the series covers a range of content from professional to academic. Typical topics might include:

- A timely report of state-of-the art analytical techniques
- A bridge between new research results, as published in journal articles, and a contextual literature review
- A snapshot of a hot or emerging topic
- A presentation of core concepts that students must understand in order to make independent contributions SpringerBriefs in Law showcase emerging theory, empirical research, and practical application in Law from a global author community.

SpringerBriefs are characterized by fast, global electronic dissemination, standard publishing contracts, standardized manuscript preparation and formatting guidelines, and expedited production schedules

Antonio Elian Lawand Junior

Legal Aspects of the Recovery of Areas Degraded by Mining in the International Seabed

Springer

Antonio Elian Lawand Junior
Charleston School of Law
Charleston, SC, USA

Maritime Law Academy
SP, Brazil

ISSN 2192-855X ISSN 2192-8568 (electronic)
SpringerBriefs in Law
ISBN 978-3-031-12491-4 ISBN 978-3-031-12492-1 (eBook)
https://doi.org/10.1007/978-3-031-12492-1

This Springer imprint is published by the registered company Springer Nature Switzerland AG
The registered company address is: Gewerbestrasse 11, 6330 Cham, Switzerland

Paper in this product is recyclable.

Foreword

Recent decades have revealed more and more the difficulty in dealing with environmental problems at both local and global levels using the legal tools traditionally used. It is necessary to create new ways of approaching and dealing with complex issues, which present themselves in very different ways from those known. This is one of the biggest challenges these days for law researchers and practitioners.

International environmental law is a recent discipline. It's essential characteristics are appropriate for the study of complex topics, such as environmental issues and conflicts in the twenty-first century. It considers, in addition to national states and international organisations, all other stakeholders involved in international issues and regimes—companies and non-governmental organisations, as well as sub-national governments, including federated states, regions, departments, municipalities, and local authorities—all capable of influencing the decisions of States and international organisations, which are subjects of classical international law.

Using more adequate and realistic tools in the understanding of global environmental issues, this new discipline allows one to be able to see all its nuances with greater objectivity and clarity. Governance among all the above-mentioned agents is the key instrument to ensuring progress in the search for viable and negotiated solutions.

This new international environmental law scenario offers researchers the opportunity to develop applied studies, with proposals of great interest. Antonio Lawand presents in this book innovative work, explaining the complexity involving the object of his research. It demonstrates that classical international law is capable of solving many issues, but in certain cases, it is necessary to go further, seeking new ways of dealing with complex problems, such as the recovery of areas degraded by mining on the international seabed, here under legal approach. The topic is current and very relevant.

Seabed mining is already a reality. Due the expansion of demand for electronic components in the world and the fact that the minerals needed for their production are abundantly found in the seabed, this new market has projected revenues in the order of trillions of dollars, which means that the trend is for this activity to intensify.

Despite market data, the normative structures that regulate mining in the Area still find gaps and apparent antinomies with other sources of the international legal regime of the Law of the Sea and other international regimes. Hence, there is a need to face the challenges that this research proposes and seek possible answers to the questions raised.

To support his arguments, the author used Case 17 of the International Tribunal for the Law of the Sea (ITLOS), which is a landmark case on the matter. The investigation of its bases and sources is of great importance and is useful for the understanding of international environmental law applied to the market for products of seabed mining.

Although Case 17 is an advisory decision, it has the merit of bringing together the first (and only) unanimous decision of the INTERNATIONAL TRIBUNAL FOR THE LAW OF THE SEA (ITLOS), addressing issues (a) of a normative nature, used in an innovative way as an integrative instrument of UNCLOS III; (b) of a doctrinal nature, represented by procedural records and their sources by the plaintiff States (Nauru and Tonga), as well as the intervening States, international organisations and non-governmental organisations (NGO).

The work deals in depth with highly complex topics of great interest. In relation to the sea, or the "Area", as the Montego Bay Convention on the Sea calls this space, the focus is on the seabed, the scientific knowledge of which is not yet fully revealed. The exploitation of funds for the removal of ores undoubtedly causes impacts, the effects of which are also not fully understood yet.

The exploitation of the seabed causes degradation, which implies the establishment of a detailed plan for the recovery of degraded areas (PRAD). This does not guarantee, however, that recovery will actually take place. This point is crucial, especially from a legal point of view, because, as the author indicates, there are legal gaps (1) in the application of the methodology for the recovery of degraded areas (RAD) to cases of exploitation of the Area and (2) in the normative references for the obligation to recover a portion of the mining area (related to the concept of "damage" in international norms).

From there, the author presents his hypothesis: "in view of the state-of-art in the recovery of degraded areas, the scientific knowledge of the seabed, and the legal nature of this geography, the currently practised methods of RAD in mining are not legally feasible. application in the Area". In line with this, the author presents a background on the identified problems. First, he mentions the existence of three types of responsibilities: those of States Parties, Sponsoring States, and sponsored companies. Everyone has responsibility for the recovery of an area degraded by mining, even if the degradation does not result from damage, acts of an unlawful nature, or omission. Second, the author points out the technical-scientific impossibility of drawing a baseline to support the PRAD in international seabeds. In other words, the main instrument to protect the explored this space is unfeasible, which exposes the risk of exploitation as it has been done.

Finally, the author discusses the need for international inter-regimental integration as a way to provide a response that does not empty the sense of the Law of the

Sea, international environmental and public international regimes in the mining of international seabeds.

As a way to give a sounded ground to his research, the author, when starting to discuss the apparent antinomy of the ISA Mining Code with the norms of international environmental law inherent to the MontegoBay/82 regime and other tangent regimes, makes use of the analysis of cases of international courts that influence the study of the precedent Case 17 ITLOS, and parallel and residually, by studies on the effects of international legal norms.

The acuity which the author approaches the matter draws attention, demonstrating that although three branches of law are applicable to the case—Law of the Sea, international environmental law, and international public law—there is still a lack of an answer that guarantees the necessary protection to the seabed. Reading the work is necessary as it not only exposes current facts but also makes pertinent reflections on the need to advance legal treatment of the matter.

Catholic University of Santos,
São Paulo, Brazil
March 2022

Maria Luiza Machado Granziera

Acknowledgements

I am honoured to express my gratitude to my thesis supervisor, Prof. Dr. Maria Luiza Machado Granziera, for her continued support and encouragement to build a deep seabed research and write down the outcomes of it. I offer my sincere appreciation for the learning opportunities provided by my supervisor.

The completion of this project would not have been accomplished without the intellectual support of Prof. Dr. Fernando Rei. Thank you for making my vision reach further.

I also express my gratitude to professors Dr. Daniel Freire e Almeida and Prof. Dr. Eliane M. Octaviano Martins and to my classmates Prof. Dr. Francisco Costa and Prof. Rosilandy Lapa.

To my kids, mom, dad, brothers, and true friends (Ana Paula, Cecília, "The Zombies", Luiz Philipe, Marina, Maria Cristina), sorry for my absence and for being grumpy sometimes. You were by my side when times get as dark as the deep seabed. To you, my gratitude is as deep as the Challenger Deep.

The present work was carried out with the support of the Coordination for the Improvement of Higher Education Personnel—Brazil (CAPES)—Financing Code 001.

Contents

Acronyms

ABNT	Brazilian Association for Technical Standards (*Associação Brasileira de Normas Técnicas*)
BFCJ	Brazilian Supreme Federal Court
BN	Brazilian Navy
BSJT	Brazilian Superior Court of Justice
CBD	1992 Convention on Biological Diversity
CBDR	*Common but Differentiated Responsibilities* or Common but Differentiated Responsibilities
CCFZ	Clarion-Clipperton Fracture Zone
CF/88	Constitution of the Federative Republic of Brazil promulgated on 5 October 1988
CHM	*Common Heritage of Mankind* or Common Heritage of Mankind
CLCS	Continental Platform Limits Commission
CRAMRA	Convention on the Regulation of Antarctic Mineral Resource Activities
GPGs	Global public goods
ICJ	*International Court of Justice* or International Court of Justice
ISA	International Seabed Authority
ITLOS	International Tribunal for the Law of the Sea
MRE	Ministry of Foreign Affairs
RDA	Recovery of degraded areas
RPDA	Recovery Plan for Degraded Areas
SACT	*Seabed Arms Control Treaty* or Treaty on the Prohibition of the Placement of Nuclear Weapons and Other Weapons of Mass Destruction on the Seabed and the Ocean Floor and Its Subsoil
UN	United Nations organisation

UNCLOS I	The Geneva Conventions of 1958, cited together with, namely, the Convention on Territorial Sea and Contiguous Zone, Convention on the High Seas, Convention on Fisheries and Conservation of Living Resources of the Sea, Convention on the Continental Shelf, and Protocol on the Dispute Resolution Compulsory
UNCLOS III	1982 United Nations Convention on the Law of the Sea
UNGA	United Nations General Assembly

Chapter 1
Introduction

The main purpose of this book is to analyse, from a juridical point of view, the recovery of areas degraded by mining in the international seabed, also statutorily called the Area (United Nations Convention on the Law of the Sea (UNCLOS) III, part I, article 1).

The investigation proposed here proves itself relevant based on several facts as stated below.

First, mining on the seabed is a technical (*i.e.* it is possible to extract ores, especially polymetallic nodules, and clusters from the seabed) and legal reality (UNCLOS III, part XI).

Second, international seabed mining is part of a new ocean market with projected revenues of around US $1.5 trillion (Organisation for Economic Co-operation and Development (OECD) 2016), given that the expansion of electronic components depends on minerals that are abundantly found in the seabed.

Third, and despite the first and second reasons, the legal structures that regulate mining in the Area still find gaps and apparent antinomies with other sources of the international legal regime of the Law of the Sea and other international regimes (such as environmental regime), which should be considered.

Fourth, the geopolitical and economic race to make sovereignties less and less dependent on important commodities from other sovereignties.

The theme of the book was chosen by inspiration on the Case 17 International Tribunal for the Law of the Sea (ITLOS), a milestone for the Law of the Sea and the investigation of its bases and sources is of paramount importance to understand the international environmental law applied to this significant new market of seabed mining.

However, the consultative decision has historically brought together the first—and only—unanimous decision of ITLOS, in addition to being the retirement session of Judge Pres. Tullio Treves from the capital office of the Court and counting on the important participation of the Brazilian Judge Marotta Rangel, the very construction of this decision (which did not result from litigation, but from consultation) was based on the following grounds: (a) normative, used in an innovative way as an

A. E. Lawand Junior, *Legal Aspects of the Recovery of Areas Degraded by Mining in the International Seabed*, SpringerBriefs in Law,
https://doi.org/10.1007/978-3-031-12492-1_1

integrative instrument of UNCLOS III. For example, the Declaration of Rio/92, an instrument, and source of international environmental law with a non-binding characteristic (so-called soft law) was taken as an integrative element and capable of binding certain obligations of Sponsor States of mining activities in the Funds of International Marine.

(b) doctrinal, represented by the procedural registers and their sources by the pleading states (Nauru and Tonga), as well as the states, international organisations, and intervening non-governmental organisations (NGOs), which conceptually increase structures not yet fully clarified by the International Seabed Authority (ISA) administrative compendiums of the precepts of the ISA regime.

In addition, our legal concern on the subject also derives from the last frontier of knowledge and occupation of civilization, which is currently the ocean. As such, both science and technology advance this geography, occupying and accessing it in the context of academic research and capitalist production or regimes of defence and international power. We will not observe this second—not, at least, directly).

But as for the arrival of the productive process in the oceans, we have mineral exploration in seabed and other uses (such as the launching of pipelines and cables or the extraction of oil from the subsoil of seabed). What matters to us, at first, is geography: if this exploration takes place in the international seabed, its legal discipline takes on special colours (given by UNCLOS III, also called Montego Bay/82, and its international regime of the Law of the Sea).

The main objective of this work is to observe the existence, the possibility, and the importance of the legal requirement for the recovery of degraded area (RDA) in international seabed when the mining event.

In order to achieve the main objective, we also observe three secondary or intermediate objectives, such as the identification of (a) the principles of international environmental law applicable to the recovery of the Area degraded by mining, (b) the legal gaps in the application of the RDA methodology to cases of exploitation in the Area and (c) normative references for the obligation to recover a portion of the area subject to mining (related to the concept of "damage" in international standards).

Our hypothesis: the scientific knowledge of the seabed, and the legal nature of this geography is that the methods currently practised for RDA in mining are not legally feasible for application in the Area. In other words, international environmental law and public international law (which are at the intersection of UNCLOS III) do not have an answer to this apparent antinomy, when exposed to an imminent or inevitable mineral exploitation.

This is because every RDA has a plan (RPDA), which, in its turn, has a baseline. A baseline has the knowledge of the natural systems (biome and ecosystem) inherent to the geography of its application. Regarding the existing geography of the Area, according to recent bibliographic surveys and the state-of-the-art of knowledge of this biome, it is recognized that we know more about the Moon than we know about the Area (Taylor and Roterman 2017, p. 4890), that is, without considering aspects of cultural and archaeological (and even ethnographic) inheritance sheltered by the ocean floor.

The Area is not only the biggest biome and the largest politically defined polygonal in Earth's crust but is also the most unknown among all the others (Taylor and Roterman 2017, pp. 4872–4879).

The imperative deduction is that the knowledge necessary to establish an adequate and reliable baseline for an RPDA proposition does not exist, making RDA an unenforceable obligation in the Area.

Our results imply the need for

(a) The identification of a set of responsibilities for the States Parties, Sponsor States and sponsored companies necessarily indicating the existence of an action to recover degraded areas through mining, even though this degradation is not the result of damage, omission, or act of an illicit (non-legal) nature
(b) The identification of technical and scientific impossibilities of drawing a baseline for the purpose of preparing a recovery plan for degraded areas (including mining areas) on the international seabed.
(c) International interregional integration, as a way of giving an answer that does not make the Law of the Sea, environmental international, and public international regimes meaningless in its cornerstone of mining the international seabed.

With these three, we started to discuss the apparent antinomy between the ISA Mining Code and the rules of international environmental law inherent in the Montego Bay/82 regime and inherent other tangent regimes.

This discussion is fueled, especially and predominantly, by the analysis of cases of international courts that influence the study of the preceding Case 17 ITLOS, in parallel and residually, by studies on the effects of international legal norms.

Thus, we divide our chapters as follows.

In the second chapter, we discuss what the Area is infrastructurally and describe and critically analyse its legal regime, so we conclude that there is an obligation for RDA on the part of Sponsoring States and sponsored entities when mining in the Area.

In the third chapter, we describe the RDA and the technical impossibilities of its realisation on the seabed (considering the current state of the art), relating it to the instruments of the Mining Code and other obligations (arising, especially, from the Case description. 17 ITLOS).

In the fourth chapter, we explore this non-response to the obligation of RDA in the light of three regimes (Law of the Sea, international environmental law, and international public law): origins and possible solutions, to work the risk and apparent breach of this international normative system (due to its normative inconsistency).

Finally, we conclude, in the last chapter, that the apparent antinomy has peculiar characteristics and is related to international environmental law and public international law, so we will discuss this subject in the proper moment, with the proper argument.

To achieve the proposed objective, we work on a structuralist theoretical guideline using the hypothetical-deductive method, as well as the data treatment methodology, obtained from selected bibliographies and documents collected from international, non-governmental organisations, national and international courts.

The methodology used was the search for bibliographic and documentary bases, being, only for the introductory descriptive purpose, research, and negligible field notes for the result.

Chapter 2
About the Area

This book focus its questions on the limit of the contemporary discussion about international law (Scovazzi 2000, p. 53): what is the regulation of international law and international environmental law for the recovery of the Area when it is degraded by mining? And it is hypothesised that there is no precise regulation on this field, or if there is one, it lacks applicability due to the lack of infrastructural/material bases/ sources. In other words, the intersection between the Law of the Sea, international environmental law, and public international law, which is the mining of the Area, does not respond to the very demands (especially, of an environmental nature) of this activity.

In terms of plan, we clarify: the Area (with a capital letter and presented in the singular) is the name by which the United Nations Convention on the Law of the Sea (UNCLOS III, in Brazil ratified by Decree No 99.165 of 12 March 1990) calls the biggest political geographical region on Earth Crust, and it "means the seabed, the seabed, and its subsoil beyond the limits of national jurisdiction" (UNCLOS II, article 1). This is the geographical limit of this research.

It is not a small epistemological cut, in fact, and so far, the Area treats just under 260 million km^2 (ISA 2010, pp. 16–19), constituting itself as the largest biome of the world (Taylor and Roterman 2017, p. 4872) in the earth's crust, with numerous ecosystems and eco-physiognomies still poorly explored (considering their wealth and size) by the infrastructural sciences (Hekinian 2014, pp. 53–54; Van Dover et al. 2016, pp. 404–405; Taylor and Roterman 2017, pp. 4872–4873 and pp. 4890–4891), with a differential characteristic of the emerged biomes: the intensity with which biotic and abiotic vectors of marine biomes operate they extend both horizontally in the landscape of the seabed and vertically in the water depth.

Biome is a scientific discretion that defines an area that brings together the same climatic characteristics and geological history associated with biotic elements common in interaction. It is called discretion, or arbitration, because although scientists justify the existence of a biome (and, especially, what makes it a different biome from another), they do it according to variables that they fundamentally value as

A. E. Lawand Junior, *Legal Aspects of the Recovery of Areas Degraded by Mining in the International Seabed*, SpringerBriefs in Law,
https://doi.org/10.1007/978-3-031-12492-1_2

important. Thus, although the biome definition is open and the definition of a biome follows the same pattern (so much so that the epistemological community diverges on how many biomes there are in the biosphere), we will define the seabed as a biome along the same lines and for the same reasons as in the bibliographies in infrastructural sciences that nourished this work.

The data provided must be weighted. Considering that, the definition of the Area in article 1, Part I of UNCLOS III based on residual determination, *i.e.*, what is not a continental platform or legally spatialised portions of the water depth (territorial sea, contiguous zone, exclusive economic zone, high seas, and/or international waters) will be the Area. Also, considering that the Continental Shelf Limits Commission has numerous requests to increase the size of this sovereign spatiality.

Finally, considering the latest consolidation of data about the Area dimension by the competent authority of its management (UNCLOS III, articles 1, 156, and 160) does not compute the continental platform extension approvals in the intervening 10 years (*e.g.* Brazil had an extension of 960 thousand km^2 of continental platform extension to the Continental Platform Limits Commission (CLCS) by articles 76 and 77, UNCLOS III, through the processes CLCS/42, CLCS/90, CLCS/103, and CLCS/108, with 85 other active submissions) (ISA 2010, pp. 16–18; CLCS 2019). Which, necessarily, makes the data out of date, but not entirely inaccurate, for the purposes it is intended in this work.

Once your horizontal limits are defined, we move on to the vertical ones: your first limit in the vertical direction, in the vector *ad caelum*, is defined by article 86, UNCLOS III, which delimits the high seas.

The opposite vector limits (towards the earth's core) include the subsoil, and from here we have a first question: What is the geological limit of the Area?

Law authors are very emphatic (and laconic) in saying it is "the subsoil" (Marotta Rangel 2014, pp. 482; Zou 2018, pp. 185–186; Wolfrum 2012, pp. 2–4), but the answer is not that simple as the sources of (international) Law of the Sea has greatly expanded the concept.

Case 17 of the International Tribunal for the Law of the Sea (ITLOS) (which will be widely referenced in the course of this work) points out the quantity and extent of activities on the international seabed (*i.e.* the Area). One possible activity is the extraction of oil and gas in super-deep waters (UNCLOS III, article 133 (a)).

The international seabed (a.k.a. the Area) is a term defined in article 1, paragraph 1 (3), of the Convention as "all activities of exploration for, and exploitation of, the resources of the Area". According to article 133 (a) of the Convention, for the purposes of Part XI, the term "resources" means "all solid, liquid or gaseous mineral resources in situ in the Area at or beneath the seabed, including polymetallic nodules". The two definitions, however, do not indicate what is meant by "exploration" and "exploitation". It is important to note that according to article 133 (b), "resources, when recovered from the Area, are referred to as 'minerals'".

Although the *International Seabed Authority* (ISA), international organisation created by UNCLOS III with the purpose of optimising, in the name of the sovereignty adhering to the treaty, the management of the Area's wealth, does not have an expressive normative structure, such as the Area Mining Code, for the exploration of hydrocarbons, does not mean that this activity is far from regulation or its powers.

We will deal with ISA below, whose missions and powers derive from Section 4, Part XI, UNCLOS III. Regarding the regulations regarding the exploitation of hydrocarbons in the Area, we have the order of Case 17 ITLOS, based on article 133 (a) and (b), UNCLOS III, for the regulation of all "minerals", including hydrocarbons, by nominal binding.

Therefore, the geological limits of the Area are defined solely and exclusively by the technical capacity of the exploration limits towards the centre of the earth, although we have never been more than half the thickness of the earth's crust, which in total is about 40 km, but we are no more than 13 km excavated.

In fact, the theme is an intersection between the Law of the Sea and international environmental law, as well as a necessary result of the material bases of both.

To assess this hypothesis and its results (be it positive or negative), it is necessary to understand the basic concepts, which are: the matrix of law in which we operate, the concepts that shape the legal universe that regulates the Area, and the fundamentals of the recovery of damaged areas (RDA).

2.1 The Area Legal Regime?

Reis and Almeida (2012, pp. 54–55), show that the process of conquering the seas evolves with the categories of use of the sea as support (from trade to the source of resources, from strategic space to the reserve of environmental services) and the technology applied on the marine theatre of operations as a process of interaction between maritime actors. At this point, it is observed that the ocean was, in international geopolitics, a medium, a theatre, and a political objective, even by analysing the police positions of the United Kingdom and the United States in the Sections XVIII–XX and XX–XXI, respectively (França 2012, pp. 76–77).

Nowadays, technology has expanded the possibilities of multi-level occupation, in the three and other categories mentioned, in different areas of the oceanic water layer (pelagic, bathypelagic areas, bottoms, etc.). With that, humanity started to perceive the ocean as a source of a greater number of riches and expanded the possibilities of this theatre of operations.

In this way, the oceanic space requires presence, strategy, and discipline that privilege one or another prevailing (sovereign) force, according to the technology/scientific or landscape advantage that this force has and its objectives. The composition among international actors in the spatial discipline of the Sea ended up having a contemporary framework at UNCLOS III (Barbosa Júnior 2012, pp. 230–231).

So, what does the oceanic "space" mean, as a theatre of operations, for the forces that contend in geopolitics and, consequently, form the law?

In the past, the seabed was a "separate planet" – scientifically unknown and impossible for human beings to occupy and perform activities on. Although it is certain to be the object of surveys, which are confused with the very birth of oceanography, since the Section XIX (Hekinian 2014, p. 2), there was no sufficient knowledge or technical capacity for exploration or occupation at this time. Therefore, as such geography was strange or inappropriate to human beings, the ocean floor was, consequently, unworthy of being a national and international normative object of the Law of the Sea.

Everything changed in the twentieth century, when humanity developed technologically to allow the occupation and activities of the last frontier of the earth's crust, the seabed. In view of this, these beds began to provoke the interest of the International Society and established themselves on the agenda when the mineral riches of which it was born were revealed: in addition to hydrocarbons in deep geological substrate, there were immersed pedological horizons, under very high pressures with riches and, among them, immense and encrusted polymetallic nodules of attractive commercial value (Hekinian 2014, pp. 162–163).

The concept of International Society that underlies the work is described by Truyol y Serra (1965), as the universalist society of sovereign nations, being as such described by the configuration of post-World War II nation states, in a universalist concept and by the self-determination of the peoples that form it. The concept is notably influenced by the legal interpretation of international normative frameworks Res. 1803 (XVII), of 12/14/1962, of the United Nations General Assembly (UNGA), the first codification of the Law of the Sea (Geneva/1958 and Geneva/1960), and international regulations on diplomatic relations (Vienna/1961 and Vienna/1963) (Truyol y Serra 1965, pp. 87–179).

Thus, there was an urgent need for the International Society to turn its eyes to this new reality of the sea and create parameters for the activities of this new theatre of operations (*i.e.* the relevant international law).

How do international law and international environmental law apply in this area?

As previously explained, the Area is a spatially residual legal concept: what is not a continental shelf (UNCLOS III, article 76), nor the high seas (UNCLOS III, article 86), and what technology can do reaching towards the centre of the earth, starting from an improper perpendicular to the international seabed, is the Area (UNCLOS III, article 1).

Regarding this space, nominally, only UNCLOS III and Geneva/58–60 (UNCLOS I) manifested themselves as framework norms. However, Costa de Souza (2019, pp. 42–43) properly indicates, which was somewhat corroborated by Case 17 ITLOS, that the Area's legal regime is an integration between inherent and peculiar legal aspects added to the regime offshore legal system and its functions.

Therefore, we must analyse both geographical circumstances to legally characterise the Area, either for the reason presented here—the rules of the high seas are integrated with the rules of the Area—or because the duality between the legal nature of both is a useful tool for the definition of legal regime.

2.1.1 The Montego Bay Convention/82 and the Governing Principles of the Area

The high seas can be legally defined by the oceanic water layer besides the exclusive economic zone, the territorial sea, or the internal waters of a State or archipelagic waters of an archipelagic State (UNCLOS III, articles 86 and 89).

Freedom of navigation and non-sovereign ownership are attributes inherent to the high seas. Scovazzi (2000, pp. 55–117), during a historical analysis, and Costa de Souza (2019, pp. 43–45), in a comparative analysis of doctrines, pointed out that this freedom of navigation of the high seas implies the use of the high seas and its potential, as provided for in article 87, paragraph 1, UNCLOS III, without having exclusive appropriation or sovereignty over it (UNCLOS III, article 89). This allows different nations (UNCLOS III, article 87) equal use of the high seas. For the same reason of permissiveness of freedom in the seas, this use must stick to peaceful purposes (UNCLOS III, article 88).

The discipline of the high seas imposed by the rules of UNCLOS III is entirely derived from this principle of freedom on the high seas and is divided into General Provisions (or Section 1, Part VII, ranging from articles 90 to 115) and Provisions of Conservation of Management of Living Sea Resources (or Section 2, which goes from articles 116–120). The importance of Section 2 lies in the recognition of the sea as common and, as such, its excessive or inadequate exploitation renders article 89, UNCLOS III, for the economic principle of appropriation. This phenomenon will be more fully explored in a later chapter. At the same time, its non-exploitation and non-occupation—in addition to directly contradicting any point of view from which the history of the human being is viewed—would make provisions for the management of resources and use, and even its very nature as common, meaningless.

Although the concept of commons finds correspondence in legally old institutes (such as the compassionate one), it was Hardin who characterised the environment as a global commons par excellence and pointed out, by criterion and game theory, the "Tragedy" that the free commons administration. Commons, in this case, is the common heritage, of collective use and of responsibility for collective maintenance.[1]

It is these two jointly interpreted sections that lead Menezes (2015, pp. 20–21) and Scovazzi (2000, pp. 78–87), even though he does not use this very same expression and derives the concept of the *Fur Seal Arbitration/1873* case, to categorise the high seas as *res commune omnium*, which means, subject to free appropriation with the limit of not denaturing or preventing the appropriation of others.

Regarding the specific provisions for the Area subject, the international legal attention to the Area has its initial mark in the preparatory acts of UNCLOS III, through the speech of the Representative of Malta in the 22nd UNGA, Mr. Arvid Pardo, on 1 November 1967 (Scovazzi 2000, pp. 118–119). Pardo's proposal, even

[1] See: Denny and Granziera (2017, pp. 9–27).

while representing a paradigm rupture in international law, stimulated, and in no way prevented the appropriation of the seabed until then, with the proviso that this appropriation was done in the name of humanity's benefit.

In other words, the richest inherent in what would come to be called the Area and have undergone a significant change in the legal regime since the normative Resolution 2749 (XXV), of 17 December 1970, of the United Nations General Assembly, changing from *res nullius* (first come, first served) to the legal category of universal equity.

From this proposition, the acts evolved to Res. 2,749 – 22nd UNGA, as part of the preparatory acts of the UN for the discussion of the treaty that would become UNCLOS III, articles 136, 137 (2), and 145.[2]

The discussion was about the regime for the use of these international seabed:

(a) Its supreme regime of freedom on the high seas would be adopted, in a *laissez-faire* system, not exclusive to other users and potential users, whose command and control would be subject to the sovereignty (or "flags", as it is predominantly in navigation); or
(b) Would a more centralised and interventional management system be adopted, under international control?

The enormous environmental accidents that occurred at the sea regarding oil tankers between the years of 1960 and 1980 during the twentieth century, from which the empirical analysis by Octaviano Martins (2016, pp. 211–213) points to direct syntax in causes with the use of the flags system and its distortion in convenience flags, helped to choose the second management model.

This cornerstone of UNCLOS III was effective in attributing to the Area a legal regime, considering the nature that we can infer and the informative principles, different from the supra-offshore. That is why the Area has the legal nature of being a common heritage of mankind (CHM). The innovation, brought by Res. 2,749 – 22nd UNGA, as part of the preparatory acts of the UN for the discussion of the treaty that would become UNCLOS III, was incorporated by UNCLOS III (UNCLOS III, articles 136 and 137), which, in relation to its predecessors, brought innovations.

---

[2]UNCLOS Article 136: "The Area and its resources are the common heritage of mankind."

UNCLOS Article 137.2: "All rights in the resources of the Area are vested in mankind, on whose behalf the Authority shall act. These resources are not subject to alienation. The minerals recovered from the Area, however, may only be alienated in accordance with this Part and the rules, regulations and procedures of the Authority."

UNCLOS Article 145: "Necessary measures shall be taken in accordance with this Convention with respect to activities in the Area to ensure effective protection for the marine environment from harmful effects which may arise from such activities. To this end the Authority shall adopt appropriate rules, regulations and procedures for inter alia: (a) The prevention, reduction, and control of pollution and other hazards to the marine environment, including the coastline, and of interference with the ecological balance of the marine environment, particular attention being paid to the need for protection from harmful effects of such activities as drilling, dredging, excavation, disposal of waste, construction and operation or maintenance of installations, pipelines and other devices related to such activities.

2.1.1.1 Common Heritage of Mankind (CHM)

Cançado Trindade (2006, pp. 365–367) pointed out that the CHM is a non-exclusive principle of the Law of the Sea, and contains two very clear dimensions that are generally applicable:

(a) A temporal one, stemming from the word heritage, in the sense that it is about the sharing of a good or value in the society through time; *i.e.*, the generations that will come should have their share of the benefit of this good and the obligation to conserve this good equally to its predecessors, bringing closer the concept of intergenerational equity[3] and CHM
(b) Another spatial/geographic one, derived from the word commons, that all international actors (not necessarily subjects of international law) are responsible for the conservation, for the inspection of this conservation, and for the extraction of benefits from these goods or values.

In other words, the author does not exclude the exploitation of the CHM, but projects it intergenerationally in equity. Considering the concept of rights and obligations, its legal phenomenology allows us to observe different obligations according to the actor of international law observed in its light. In the case of States, the same author reveals that the good or value subject to CHM will be the object of a public form of trust, with the sovereign agent being one of several trustees (according to their limits of action given by the international treaties and their own sovereignty).

That said, instead of a principle of appropriation for resumption typical of *res nullius*, the Area has as a principle the practice of supporting activities (*e.g.* economic) of a State if it does not denature other activities to other States when they can/want to exploit it (in other categories of production).

For example, the Area is qualified by Res. 2,749 – 22nd UNGA, taking the juridical nature of *res commune omnium*, or, according to Cançado Trindade (2006, p. 373), *res communis humanitatis*.

The idea of *mare liberum* (since Grotius), taken by the radical and conceptual bias, is the same: to use, to enjoy, and to allow others to do so in proportion to their need. The evolution of the concept of *mare liberum* took place, exclusively, in terms of infrastructure: (a) the need for a global development recognised by the International Society (in this case, in its most prestigious UNGA pulpit); (b) the scientific and technological advances that made it possible to get to know the Area, the acknowledgement of its riches and the possibility of exploiting them.

In this sense, we can rely on the analysis of Costa de Souza (2019, pp. 57–58), who explains the Area as a CHM, and as the legal materialisation of a demand for a new world of economic, intra- and intergenerational order.

---

[3] On the Principle of Intergenerational Equity in International Law, see WEISS, Edith Brown. Intergenerational Equity: A legal framework for global environmental change. See: Weiss (1992, pp. 385–412).

Grotius (2002, pp. 20–39) pointed out that the reasons for this very special rule in relation to the freedom of use of the sea and its resources derive from the same logic of peaceful use of the sea. It is not allowed for one nation to use the sea to the detriment of another because of two fundamental principles of international law: the principle of no harm and the peaceful use of the oceanic sea. Exhausting the shoals, for example, would not only privatise the sea in relation to a single national dominus, but also under the auspices of the freedom to sail, one country committed an abuse of this right to harm (from absolute lack of fishing product) the other (in the same bias of article 63, UNCLOS III, as interpreted by Case 21 ITLOS and, *ab origine*, the *Trail Smelter* case). With a lot of propaedeutics, he comes to compare this hypothesis with the administration of the (*millennial*) institute of the compassionate: a common good for all, necessary for the economic subsistence of all, of free access and use in the limit of not mischaracterising it and preventing the use of others (Grotius 2002, p. 87).

To be more explicit, the decision provides *ipsis litteris*:

> 198. The Tribunal is of the view that the term "development of such stocks" used in article 63, paragraph 1, of the Convention suggests that these stocks should be used as fishery resources within the framework of a sustainable fisheries management regime. This may include the exploitation of non-exploited stocks or an increase in the exploitation of under-exploited stocks through the development of responsible fisheries, as well as more effective fisheries management schemes to ensure the long-term sustainability of exploited stocks. This may also include stock restoration, guided by the requirement under article 61 of the Convention that a given stock is not endangered by overexploitation, thus preserving it as a long-term viable resource.
>
> 199. Article 63, paragraph 2, of the Convention establishes a cooperation regime between the coastal State and the States fishing for the same stocks and stocks of associated species with a view to agreeing on measures necessary for the conservation of these stocks in the adjacent area.
>
> 200. Since the Tribunal has jurisdiction to entertain the Request only in so far as it relates to the exclusive economic zones of the SRFC Member States, article 63, paragraph 2, of the Convention, as far as it relates to "States fishing for such stocks in the adjacent area", is not applicable to the exclusive economic zones of the SRFC Member States" (ITLOS 2015).

To explain how ITLOS CASE 21 started and what its object is, the following is a brief discussion: on 28 March 2013, the International Tribunal for the Law of the Sea received a request from the Sub-regional Fisheries Commission (SRFC) to process an Advisory Opinion.

SRFC is in Dakar, Senegal, and is composed of seven Member States: Cape Verde, Gambia, Guinea, Guinea-Bissau, Mauritania, Senegal, and Sierra Leone. In a resolution adopted during its 14th extraordinary session (25–29 March 2013), the SRFC Conference of Ministers authorised the SRFC Secretary "to take over the International Tribunal for the Law of the Sea [. . .] in total their advisory opinion on the following matters:

1. What are the obligations of the flag State in cases where illegal, unreported, and unregulated fishing activities (IUU) are conducted within the exclusive Economic Zone of third States?

2. To what extent will the flag State be held responsible for IUU fishing activities conducted by vessels sailing under its flag?
3. When a fishing licence is issued to a vessel under an international agreement with the flag State or an international agency, the State or international agency shall be liable for the violation of the fishing legislation of the coastal State by the vessel in question?
4. What are the coastal State's rights and obligations to ensure the sustainable management of shared stocks of common interest, especially small pelagic species and tuna?"

The resolution was adopted pursuant to article 33 of the 2012 Convention on the Determination of the Minimal Conditions for Access and Exploitation of Marine Resources Within the Maritime Areas Under the Jurisdiction of the Member States of the Sub-regional Fisheries Commission (CRFC), through which the Conference of Ministers may authorise the Permanent Secretary of the SRFC to take questions from the Court.

And, in the same context, a brief about *Trail Smelter* case: in 1941, the United States sought compensation from Canada through a lawsuit and filed for an air pollution injunction in Washington State by Trail Smelter, a Canadian company domiciled in Canada. The Smelter Trail, located in British Columbia since 1906, was owned and operated by a Canadian company. The sulphur dioxide from the smelter resulted in damage to the state of Washington between 1925 and 1937. This led to the United States filing a lawsuit against Canada.

The case discussed the duty of one State to protect other States against any harmful acts perpetrated by individuals within their jurisdiction (duty of no harm).

But is it the responsibility of States to protect other States from the harmful acts committed by individuals within their jurisdiction at all times? The answer is yes. No State has the right to use or permit the use of the territory in such a way as to cause smoke wounds in or on the territory of another or on the properties or people therein stipulated by United States law and the principles of international law.

Mutatis mutandis, the Area became common in 1970, through Res. UNGA 2749 (XXV). The moment could not be different, according to the following reasons:

(a) Scovazzi (2000, pp. 117–119) extracts from Pardo's speech, in 1967, the growing economic motivation at the time: it was part of the North-South dichotomy to make the North increasingly independent of the South by the search for autonomy in relation to the supply of rare mineral commodities. The South, on the other hand, understood that its technological shortage could not be a reason to be, once again, a supporting factor in yet another world economic cycle (*i.e.* mining in the Area) that was about to start.
(b) Oda (1969, pp. 458–460 and pp. 464–466) focuses his attention on the peaceful uses of the oceans, as a customary rule consolidated in UNCLOS I (the only law in force at the time), not dissociating it from the (then) novelty access to the area. Oda already advocated the notion that the use of the Area influenced the use of the adjacent water depth and vice versa: the use of international funds in a non-peaceful manner emptied the entire international construct of the peaceful

use of the seas (since Grotius). Such a point of view was justifiable and relevant, given that it was formulated in the middle of the Cold War, just eight years after the Cuban Missile Crisis (1962), and aware of the missile technology with potentially launchable nuclear warheads from submarines.

(c) Evensen (1986, pp. 430–434) considers the events of the New World Order (the bipolar world of the late half of the twentieth century and the seven powers the US, the EU, the UK, France, Germany, Italy, and Japan) and technological advancement as the fall of the old legal regime of the sea (UNCLOS I), saying it insufficient to the new oceanic realities.

Being common, rules of access and accountability are being structured (originating from this idea of Grotius until nowadays). In fact, the (mineral) exploration of the Area obeys a series of requirements previously checked by a delegated international authority (under a legal regime of the international organisation), under a licence regime, which observes the country's ability to explore this Area without mis-characterising it in definitive or prevent its use/access to other nations. In a timely chapter, we will define these normative criteria and compare it with the protection of the value (Area as common) to which it is proposed.

The most important thing, at the moment, is to understand the idea in Law of the Sea that a right to the utility of a common has a limit, both in the portion of the sea subject to sovereignty, and in the same right of the other State to the same common, throughout a present and future time, of a three-dimensional elasticity in space-time.

In the case of a CHM subject to the regime of the sea, *i.e.* the Area, all sovereignties operate at the same time as trustees and as stakeholders, with the administration of this trustee being legally disciplined in two phases: (a) a general one, by ISA and its administrative bodies; and (b) another when the mining phases (exploration and exploitation) are opened, in which the geography of the mining area is subject to the trust of one (or more nations in consortium), and to the contracted company (*i.e.*) under the public regime.

As for ownership: they are the same, namely humanity. But can human beings be the subject of such rights? Costa de Souza's reading (2019, p. 54) about the case of *Count Bernadotte/1949-ICJ*,[4] and Cançado Trindade (2006, pp. 365–367) in the reading of the obligations of States when faced with one or another well declared as these principles, which illustrates the figure of the trustee, leaves no doubt: the answer is positive.

Although, the subject of rights, the humanity, does not have the power of procedural representation, for the purposes of the law of the sea. Gautier (2016) and Tanaka (2018, p. 131), when analysing the same legal assets related to the ownership of humanity, both using the judgments *Southern Bluefin Tuna/2000* and *MOX Plant/2003*, pointed out that the courts and arbitration in international public law (in particular, the Law of the Sea) welcome the protection of these assets, but reserve the ability to be part to the national States (although not materially affected),

[4] About the Count Bernadotte case, see ICJ (1948).

by the principle of sovereignty. Gautier goes further, while Tanaka (2018, p. 131) harmonises with Gautier (2016) pointing out that the postulation of an uninjured country (victim country) against another offender of a heritage of humanity derived from the general violation of treaty also uses the principle and inherent question of world heritage for the condition of the sovereign State to be a party in litigation (although not directly offended). But at the same time the representative of humanity (*i.e.* the individual) considered as the subject of this right cannot be a party to an international court. Sovereignty is still a fundamental element, even if, considering the provisions, it makes its holder something close to an extraordinary legitimacy of humanity's interests.

If the ownership is of humanity, as declared by representation of their sovereign States, the same States can institute a body to administer this right, in the absence of the right of any one of them to administer unilaterally. The body was created and instituted: it is the International Seabed Authority (ISA) (UNCLOS III, article 156) that operates, through provocation by claims of States Parties, in representation to all States Parties and, in ultimately, to humanity (as "trustee" of this trustee), authorising or not authorising operations in the Area, according to the dictates of article 157, UNCLOS III, especially the CHM.

As for the content, let us point out what these legal regimes should embody about the CHM, which follows, described in the form of principles.

2.1.1.2 Governing Principles in the Area (as CHM)

For international environmental law and the Law of the Sea, the normative force of the principles (whether positive or customary) appears not only in the judgments, but in the way in which the precept and the legal norm of the Law of the Sea is constructed.

Wolfrum (2012, pp. 3–5) describes the application of CHM to international seabed as unfolded in some principles applied to the Area use regime in themes, such as a ray of white light that refracts other colours and frequency when it crosses a prism. Although using other indicative terminology, Menezes (2015, pp. 150–156) points out the same principles refracted as autonomous and at the same level as the CHM, in view of their geography in UNCLOS III.

In fact, the consequences of one or another notion do not affect the implementation of the principles, as they are considered in the light of the post-positivist concept. In this way, we pass on to them.

Obligation to Ensure Compliance with the Law and Liability for Damages

Regarding the obligation to ensure compliance with the law and liability for damages, Menezes (2015, pp. 151–152) describes a principle inculcating this duty in States. They must ensure that activities in the area, conducted by them, state-owned companies or by legal entities, natural or legal, which hold the nationality of the

States Parties or are under the effective control of those States or their nationals, or organisations international, are conducted in accordance with the Convention.

An important theme for international law is that the Convention ends up disciplining the international responsibility of the State in case of non-observance of the provisions of the Convention, as well as solidarity when they act alongside international organisations, not being extended to the State when sponsoring a legal person.

To protect a CHM, it is necessary to have caretakers (*i.e.* responsible) and their delegates in action in favour of a type of quality to be maintained in the Area.

To explain the topic, we return to Case 17 ITLOS, which states that it is the responsibility of States to repair environmental damage incurred in the Area, except if measures of a contractual, administrative, and legal nature are conducted within the countries sponsoring activities in the Area, in accordance with the mining code in the Area. That is to say: depending on whether the harmful event fits the moulds and legal limits of liability created by UNCLOS III, Mining Code ISA, and Case 17 ITLOS, private entities that explore and exploit the Area and/or their Sponsoring States. These criteria will be discussed in a specific chapter, with a view to ensuring responsibility for recovering degraded areas.

But what about other activities? There is no point in saying that other activities are outside of international regulation for three reasons.

The first, for treating the Area as something common to all and, as such, subject to the observance and protection of all, as mentioned above. An attack on a common of this nature is an attack on the whole of humanity, its natural, biotic, and abiotic reserves of resources, subject to the legitimate use of force by members of the international community. This conclusion is obtained from an analysis of two cases, namely ITLOS 3 and 4—*Southern Bluefin Tuna* cases (ITLOS 1999) and ICJ—*Whaling in the Antarctic* (ICJ 2014).

In ITLOS 3 and 4—*Southern Bluefin Tuna* cases (ITLOS 1999), Australia and New Zealand filed, on 30 July 1999, requests for provisional measures with ITLOS. The requests ask the Court to prescribe provisional measures (a preliminary injunction) obliging Japan to immediately cease the unilateral trial of bluefin tuna fishing, which began in early June 1999. Japan submitted its response to the requests on 9 August 1999. In its response, Japan maintains that the Court should deny the provisional measures requested by Australia and New Zealand. It also contains a counter-request from Japan for interim measures.

The governments of Australia and New Zealand decided to submit to an arbitration procedure, under UNCLOS III, Annex VII, their dispute against Japan. Pending the establishment of an Arbitral Tribunal, the governments of Australia and New Zealand requested the International Court for the Law of the Sea to prescribe provisional measures (injunction) pursuant to paragraph 5 of article 290 of the Convention. The Court heard the parties' oral arguments in five public sessions between 18 and 20 August 1999.

Japan was ordered to cease fishing activities not included in the agreement.

On 31 May 2010, Australia filed a case in the International Court of Justice (ICJ) against Japan over Japan's "continued attitude to a large-scale whale hunting programme under the Second Phase of its Japanese Whale Research Program under a Special Licence in Antarctica (JARPA II), in violation of Japan's obligations under the International Convention for the Regulation of Whaling (ICRW), as well as its other international obligations for the preservation of marine mammals and the marine environment" (ICJ—Whaling in the Antarctic (ICJ 2010)).

As a basis for the jurisdiction of the Court, Australia invoked the provisions of Article 36, paragraph 2, of the Statute of the Court, referring to statements recognising the jurisdiction of the Court as mandatory made by Australia and Japan on 22 March 2002 and 9 July 2007, respectively.

On 20 November 2012, New Zealand filed a statement of intervention with the Secretariat. Based on Article 63, paragraph 2, of the Statute, it claimed that, as part of the International Convention for the Regulation of Whaling (ICRW), it had a direct interest in the interpretation that could be placed on the Convention by the Court in its decision in the process.

By means of Resolution of 13 February 2013, the Court, having observed that New Zealand complied with the requirements set out in the Court's Statute and Rules of Procedure, found that the statement of intervention was admissible. Public hearings were held from 26 June to 16 July 2013, during which oral arguments were presented by Australia and Japan, and the experts that each party asked to be called were heard by the Court. New Zealand made oral observations about its intervention.

In its judgment delivered on 31 March 2014, the Court first found that it had jurisdiction to hear the case, rejecting Japan's argument that the dispute fell within the scope of a reservation contained in Australia's statement recognising the jurisdiction of the Court as mandatory. Then he turned to the question of the interpretation and application of Article VIII of the 1946 Convention, paragraph 1, which states that the parties "may grant any of their [nationals] a special licence that authorises that national to kill, catch and treat whales for the purpose of scientific research".

To interpret this provision, the Court noted that even if a whale-hunting programme involves scientific research, the killing, capture, and treatment of whales in accordance with that programme does not fall under Article VIII unless those activities are "for the purpose of" scientific research. To determine this point and to determine whether the use of lethal methods by a programme is for the purpose of scientific research, the Court considered whether the elements of the design and execution of a programme were reasonable in relation to its declared scientific objectives.

With respect to the application of Article VIII, paragraph 1, the Court noted that JARPA II could be broadly characterised as "scientific research". However, the Court considered that the evidence presented did not demonstrate that the design and implementation of the programme were reasonable for the achievement of the stated objectives. The Court concluded that the special authorisations granted by Japan for the slaughter, capture, and treatment of whales in connection with JARPA II were not "for the purpose of scientific research", in accordance with Article VIII, paragraph 1, of the 1946 Convention.

The Court then turned to the implications of that conclusion, considering Australia's claim that Japan had violated several provisions of the list annexed to the Convention. Having discovered that Japan had in fact violated some of the invoked provisions (notably the moratoriums on commercial whaling and factory ships and the ban on commercial whaling in the Southern Ocean Sanctuary), the Court moved on to the issue of remedies. Since JARPA II was an ongoing programme, it ordered Japan to revoke any existing authorisation, permit, or licence to kill, capture, or treat whales in relation to it and to refrain from granting any other permit under Article VIII, paragraph 1, of the Convention in compliance with that programme.

In all three cases, they litigated in consortium Australia and New Zealand against fishing practices conducted by Japan under specific treaties, both under the rule of the Law of the Sea regime.

Japan's preliminary claim, as a passive pole, was that, regardless of the fishing qualification it practised, in no way did this fishing damage sovereign domains of Australia and New Zealand, which would render both nations deprived of the right to act in an international court (under the terms of article 290 of UNCLOS III).

In return, the reply of both States of Oceania has stipulated: the three States comprise parts of the same treaties (UNCLOS III, Convention for the Conservation of Southern Bluefin Tuna – Canberra 1993, and the Washington Conference 1946).

The treaties, in turn, seek to protect a legal good and presuppose a good faith agreement between mutually recognised sovereignties; the violation of the legal property protected by the treaty by one of its parties creates three types of violation: (a) material, by the rupture of a legal property of collective interest, insofar as it has been (con)tracted/treated and as such declared by the States Parties. This situation is completely different from the *Trail Smelter* case, which had no other legal-environmental regime to solve the question of the evolution/reconstruction of the interpretation of the Peace of Westphalia (1648); (b) formal, for the breach of the universal duty of good faith between contracting nations (Vienna Convention 1969); and (c) formal, for ignoring a sovereign decision of another nation, being, as such, a partial and indirect attack on this sovereignty (Vienna Convention 1969 cc. Peace of Westphalia 1648). *I.e.*, the violation of a Treaty whose object is a universal and/or diffuse right agreed between the parties, *per se*, generates damage to non-violating contracting parties, even if not materially consolidated in the sovereign territory itself.

Both international courts welcomed the Australian and New Zealand replicas. The material argument was especially highlighted: an international regime on a given asset makes this asset of interest to all contracting parties. The illegal rupture of this regime means the illegal rupture of the interests of all contracting parties, entitling them to the defence of these violated interests (ITLOS 1999, pp. 297–300; ICJ 2010, pp. 1–2).

Applying the concept to the species, we have that the Area is a CHM due mainly to the 22nd UNGA. This, in turn, draws strength from the United Nations Charter - San Francisco/1945, an instrument binding on all UN Member States.

As a result, it is quite clear that the violation of the Area's regulations or its depletion illegally considered is an attack on the heritage of the 195 members of the UN. And any of them, even if not a signatory to UNCLOS III, is entitled to defend it through the means of solving international conflicts.

Established that the legal or contractual violation, as is the case of the Area's mining regime—explored in the next chapter—is liable to impose the duty of re-composing (without eventually resulting in any kind of harm to others), we ask: What about the damage as the one arising from operations that are entirely legal, legitimate and within a pre-established legal criterion without violations by any parties?

From the economic infrastructural perspective, the act of polluting/degrading the Area is an act of appropriation (Derani 1997, p. 73) and, as such, prohibited by the rule literally obtained from the interpretation of article 1 (1), UNCLOS III, which would immediately make the State Party responsible for its act.

As stated before, except for mining, no other activity or heritage had its management of the Area detailed by ISA. However, human activities at sea (on the surface, in water depths, or on the international seabed themselves) invariably lead to the appropriation of these bottom areas. In 2019, explorer Victor Vescovo descended to the Challenger Deep (the deepest point of the Earth crust, in the Mariana Trench in the Pacific Ocean near Guam) and found, among biotic and abiotic resources, a plastic bag and a candy pack (Morelle 2019).

It is necessary for the purpose of this study to point out that at no instance in Part XI is there an obligation to prepare a recovery plan for damaged areas (RPDA). It is also important to mention that the management speech of the Area made by ISA (Wedding et al. 2015, p. 144) or made by UNCLOS III (Part XII) is predominantly preventive.

But mining is required, and invariably appropriates the international seabed, as recognised in the international law sources interpreted in the courts legitimised by UNCLOS III.

Hence, the normative gap is perceived by the movements of the international community: at the time of the writing of this study, within the scope of the UN, under the UNCLOS III regime, marine blade biodiversity and funds beyond national jurisdictions are discussed. The International Conference on Marine Biodiversity in Areas Beyond National Jurisdiction, already in the fourth round of discussions, but without a text contracted between States Parties, yet, by force and order of Res. 72/249, UNGA.

The Conference's mission is to develop a hard law document (binding) on the protection and management of this marine biodiversity, the legal nature of which is already found in Part XII, UNCLOS III.

Not only the law is a formal source of international law. How the sources that nourish the International Law of the Sea can structure, even though they point to principles, analogy, equity, customs, and unilateral acts of States and international organisations as a way of integrating defective norms (Cançado Trindade 2006, pp. 147–168).

Following this structure, let us observe how ITLOS treated the mining activity in what the ISA Mining Code itself was lacking: in a circumstance in which a harmful activity (other than mining) sponsored or undertaken directly by a State Party in the Area can be, as appropriate, held responsible in the same manner as the Mining Code, Part XI, UNCLOS III, and Case 17 ITLOS, through the Sponsoring State's due diligence regime to its sponsored private body.

In this case, the Trail Smelter would have its share of application: it is not given to sovereignty, even in acts of perfect formal legality, offending other sovereignties in its assets. In other words, although within the parameters indicated by the ISA, the RDA obligation when decommissioning mining or exploratory units in the Area is a mandate that does not derive from Part XI, but from the fundamental concept of multiple access CHM.

Therefore, protecting the marine environment means freeing it from damage according to the provisions of article 235, UNCLOS III.

There is more than one notion of damage that can be addressed in the definition of the principle in question, taking the point of view of differentiating responsibilities from liabilities of ITLOS (2011, p. 24): the normatively arbitrated damage and the *ex delicto*.

In a brief explanatory tour, ITLOS (2011, p. 24) defines the following terms:

> 69. Thus, in Question 1, the term "legal responsibilities and obligations" refers to primary obligations, that is, what the sponsoring States are obliged to do under the Convention.
>
> 70. In Question 2, the English term "liability" refers to the consequences of a breach of the sponsoring State's obligations.
>
> 71. In Question 3, as well as in Question 1, "responsibility" means "obligation". The terms "Responsabilité" and "responsabilidad," used respectively in the French and Spanish versions of Question 3, are translations of the English term "responsibility" and were apparently introduced for the sake of uniformity. However, in the light of the English version and the terminology used in the French and Spanish versions of Article 139 of the Convention, the intended meaning is "obligation". Likewise, the Arabic, Chinese and Russian versions of Question 3 use the term "??ل???م", "义务" and "обязательство", respectively.

Considering the judgment, international environmental law finds in the Area regime two categories of environmental damage: that resulting from the liability regime of the treaties (article 60, Vienna-I 1969) and the liability regime of the States; and those that result from a breach of obligation (called *ex delicto*), expressed in the case analysed, and those that result from sine *delicto*, *i.e.*, that happen without a formally anti-legal conduct of the agent.

Regarding *ex delicto* responsibility, it arises from an obligatory breach resulting from a treaty or custom, in the context of a regime (Fitzmaurice 2002, pp. 204–205). Although the questioning of ISA to the Specialised Chamber of ITLOS refers (in title) to the responsibility of the State, it is a fact that the content of the questioning ISA as the judgment of Case 17 ITLOS suffered an epistemological cut inherent to article 60, Vienna-I 1969, when defining responsibilities and liabilities. It should also be noted that the liability regime based on failure to comply with the treaty obligations was expressly mentioned when citing the Gabčikovo-

Nagymaros (ICJ) case. as a source to structure the liability regime enunciated by mining on the international seabed (Fitzmaurice 2002, p. 207).

However, and in this case, the offended party is any counterparty to the treaty (see Bluefin Tuna Case ITLOS), as well as, the CHM (nominal legal binding nature of the Area) case, being the benefit for humanity (see the next topic), the counterparty whose environmental assets have been harmed will be any UN Member State.

Benefit for Mankind

It is a direct command from UNGA for UNCLOS III Convention. Otherwise, is directly related to game theory applied to a common.

Gonçalves and Costa (2011, pp. 127–139) describe the unnusual approach, of a commons by its actors in an environment of decentralised or concentrated power, while Denny and Granziera (2017, pp. 15–20) do so from the perspective of "Tragedy of the Commons". In all constructions, two variables impose themselves: benefits and/or access. Whether one or the other, the quality and quantity of these variables will determine, in economic criteria, the gain of each actor in the problematic distribution of commons.

In the case of the Area, the solution was not to limit access to commons (although it is provided for in ISA regulations and manuals). But, yes, no matter who accesses, collectively or individually, a geography of the Area for mineral exploitation, this must be done considering that there will be a sharing of the benefits. In the same sense, Egede (2018, p. 158).

Following there is presented the benefit of sharing of Part XI in three-order obligations:

(i) Activities in the Area, including mining, should generate revenue for a common and international fund destined to the development of less favoured nations and the management and development of the Area itself
(ii) The reservation of portions of the Area for mining by developing countries or for the benefit of the latter in a special mining regime
(iii) (Not mandatory, after the 1994 reform), the country that sponsors mining in international funds commits to transfer technology to the developing nation so that it develops a similar activity

Costa de Souza (2019, pp. 57–58) clings to the economic content of Pardo's speech regarding the economic development of nations, especially the poorest. That is, the author treats CHM as a tool for the redistribution of wealth in the world according to a global ethic of strengthening the most fragile, within a developmental and trans-generational (*i.e.* sustainable) aspect. Cançado Trindade (2006, p. 366) shares the same opinion.

Use of the Area Exclusively for Peaceful Purposes

We cannot forget that this unfolding of the CHM in the Area derives both from the evolution of Grotius' postulates about the sea and its uses (Scovazzi 2000, pp. 76–87), as well as from the historical (geopolitical) moment in which Tommy Koh captained the articulation to the achievement of the concepts discussed here in an international codification of the Law of the Sea: between the years of 1972 and 1982 there were two identifiable axes of geopolitical conflict, both incident on UNCLOS III matters, namely, the East-West conflict axis, also known as "Cold War", and the North-South conflict axis, which dealt with the depletion of natural resources *vs* the right to development.

Both variables were significant weight in the use of the Area exclusively for peaceful purposes, given the geopolitical role that determines the conquest of the seas in history (Mackinder 2004, pp. 293–321), and the continuous need for scientific advancement over the domains of the sea (Menezes 2015, pp. 153–154).

The option to qualify the Area as CHM can only be the appropriation of its specific geographies with peaceful motivation and with the promotion of scientific development or, at least, preservationist (in view of the trans-generational purpose).

Likewise, the peaceful use of the seas is a general concept of the sea regime as denoted in the *Fur Sea Arbitration*/1895 (Scovazzi 2000, p. 87). Thus, it is logical that, for this principle to function properly, all political divisions of the sea, as defined in UNCLOS III, are contemplated with peaceful use as well. Hence, the peaceful use of the seas contains the principle of peaceful use of the Area, in addition to deriving it.

The question is: What defines this peaceful use?

This question is important, the Area supports a series of scientific activities and/or technological applications that serve both peace and war activities (*e.g.* submarine cables). Likewise, the various politically defined portions of the sea (including the Area) suffer, as never before, pressure on their conflicting uses, which can denote in situations bordering on belligerent acts (Myrdal 1971, p. 6) (as in the Horn of Africa and the South China Sea).

The critical analysis of the ICJ *Affaires Nucleaires*/1974 case, which opposed Australia and New Zealand to France, due to the tests of nuclear artefacts and their effects (especially the fallout) in the Bikini Atoll (Pacific Ocean) brought up the discussion of what peaceful use is, and whether it is antagonistic to military use or imposes limits on scientific use.

Fallout or radioactive precipitation effect "is the effect on the ground of residual radiation from the large amount of radioactive debris produced by nuclear detonation that goes up to the atmosphere and is deposited later. (...) Radioactive precipitation has the following characteristics: a) the amount of material contained in the radioactive precipitation depends on the power, the type of weapon and the height of its detonation; b) the area affected by the contamination of radioactive precipitation depends on the direction and speed of the wind [in the case of the Area, the currents]; c) surface detonations are the ones that most produce radioactive precipitation; d) being directly linked to radioactive precipitation, the Radiological Hazard arising from nuclear detonation depends on the level of the radiation dose rate present and the exposure time of the victims; and e) climatic conditions considerably affect the extent of the affected area" (Brasil 2016).

The answer came from Myrdal (1971, pp. 6–9 and 11), who concluded that peaceful activities are not synonymous to civil activities nor the antonym of military activities. In fact, the author makes the syntax of the peaceful use of the Area with activities capable of causing nuisance, or, in her words, an impact outside the "reasonable", in a very typical framework of the duty of no harm, when viewed in the light of an asset common to all jurisdictions (such as CHM).

Thus, a civil activity will be non-peaceful when it creates losses or hindrances to the right of access to the Area of other sovereignty in an intentional or culpable way, without due observance or risk management (this, yes, deliberately). The reverse is also true: the transit of a military machine in the middle of the Area, even if they are experiments for the purpose of future military armament, can be considered "peaceful use", in this interpretive tuning fork.

Although the ICJ *Affaires Nucleaires*/1974 was finalised by the loss of the object, the Australian and New Zealand argument was oriented in the same direction, with New Zealand trying to reopen the case recently under it (ICJ 2015, pp. 94–99). For the same reasons, motivated by the relevant Study and Writing Committee, the Treaty on the Prohibition of the Placement of Nuclear Weapons and Other Weapons of Mass Destruction on the Seabed and the Ocean Floor and Its Subsoil (SACT), 1970, prohibits nuclear weapons and mass destruction weapons, in any capacity (warlike or scientific use), on the seabed and adjacent deep sea in addition to the 12m nautical off the coast of the signatory sovereignties (Myrdal 1971, pp. 11–12).[5]

In this way, mining on the international seabed will conform to the principle in question as long as: (a) it is not used as a weapon of war, that is, it will fully harm a sovereignty or the patrimony of all; (b) is not conducted in a disrespectful manner with respect to environmental responsibilities, whether deliberately or not; or (c) do not disturb the peace and property of others outside of the treated exercises, even if the operation does not have this intent.

Protection of the Marine Environment

Menezes (2015, p. 154) states:

> When using the area, the maritime environment must be protected and preserved in the perspective that it must be a space for maintaining life, essential for the marine and terrestrial ecosystem. States should prioritize the use of the area for the protection of the maritime environment against the harmful effects that may result from the activities developed in its use.

[5]The French Republic is not a signatory to the said Treaty. Treaty was signed by Brazil, ratified and promulgated in the internal order through Decree No 97.211, on December 12th, 1988.

> This principle involves the commitment of States in the use of the area, that is, the concern with preventing, reducing and controlling pollution and other hazards to the maritime environment, including the coast, as well as disturbing the ecological balance of the maritime environment, avoiding drilling, dredging, excavation, dumping of debris, construction and operation or maintenance of facilities, pipelines; it also involves the protection and conservation of the area's natural resources, preventing damage to the flora and fauna of the maritime environment.

The definition above allows several inroads. First, activities in the Area must not interfere negatively with the marine biome (*i.e.* the syntax between marine ecosystems adjacent—vertically and horizontally—to the portion of the Area in which the impact occurs). It deals with the present and future potential uses of the various States preserved through the utilitarian view of the marine environment (Barboza 1994, p. 397).

As a way of being more detailed, Case 17 ITLOS (2011, pp. 27–31) starts from the notion of the exploration and exploration process of the Area (called "Activities in the Area"), implies admitting activities related to mining that geographies other than international soil and marine subsoil are used, such as the water column (and its various oceanographic divisions), the surface and coastal areas.

Second, the rule derives from the notion of the intrinsic value of the environment, expressed in Part XII, UNCLOS III (Barboza 1994, p. 397). The environment, in UNCLOS III, is taken today as "common concern".

Despite UNCLOS III and its Part XII having been drafted and edited under the aegis of the Stockholm Declaration/1972, the case WTO 58/1998 "Shrimp-Turtle", which opposed India *et al.* to the United States, it is sufficiently illustrative to state that the norm that derives from the precept of the treaty evolves over time and nourishes itself with information from the new times for this evolution.

Thus, taking the duties of observance of the responsibilities and related to the environmental damage in Case 17 ITLOS, we have that the protection of the maritime environment in the Area has three distinct moments, which can be guided by the Principles of the Declaration of Rio/1992: the preparatory activity in the Area, notably regulated by the precautionary principle; the course of activities in the Area, with control and compliance measures in place, also regulated by the precautionary principle, in the context of prevention; the occurrence of the damage, which may be due to a mandatory breach, the first moment of which requires the containment of this damage, notably regulated by the polluter-pays principle; and environmental recovery, as a way to make the marine environment restored and less vulnerable in degraded portions and, with that, resilient.

Note that the principle in question has a preventive burden ("prevent, reduce and control") much more intense than that of repair.

Therefore, the first hypothesis arose from the term "control" in article 145 (a), UNCLOS III. The article, if interpreted according to the technological limitations of exploration of the Area and its multiple uses, denotes that RDA would be an instrument to prevent the spread of new damage to the environment from international funds, just like a firebreak as a defensive measure against forest fires. In this case, and as an instrument, it would be up to science, and not to law, the binding opinion of its application in the event of damage. This is because RDA, in this case, would become an alternative among many, with the law being only the result of caution/control of ecosystem risks.

The second is that the protection of the marine environment also includes repairing the damage, as a necessary derivation of article 235 cc. 145 (a), UNCLOS III, when forced to give mandatory legal meaning to the word "reduction".

Reducing pollution and/or degradation has its preventive character, that is, reducing the load that the production unit emits on the environment, as it has its remediation face, that is, recovering degraded areas.

In this second sense, protecting the marine environment means increasing its original characteristics, allowing it to reach the closest to the status quo in the face of degradation.

Protection of Human Life

The principle that brings together an intersection between various regimes is the protection of human life at sea, and by the sea is governed not only by the legal dictates of the Area, but also by the international navigation regime, of the Universal Declaration of Human Rights (UDHR), of International Labour Law/MLC/2006, the International Convention for the Prevention of Pollution from Ships (MARPOL), among others.

Considering the human participation in the mining processes of the national and international seabed, this principle has its application frequently elevated in the exploration of the Area, especially in the incursions with submersibles to the environments of high pressure. The epipelagic operations (and even the beginning of the mesopelagic) of the water slide, which occur both in exploration and exploitation, are guided by the same International Maritime Organization (IMO) criteria.

Harmonisation of Activities in the Area and in the Maritime Environment

From the stated principles, this one holds a lot of importance for the conclusions of this work regarding the harmonisation of activities in the Area and in the maritime environment.

This is identifiable from the content that defines this principle, namely:

> States should use that space, installing their structures, without radically transforming the panorama of its use and not intervening in the activities naturally developed there (Menezes 2015, p. 155).

Therefore, the principle in question implies that the present and future potentialities, as well as the services (environmental, for example) supported by the Area's ecosystem, should not be denatured by an international actor that acts in it, with or without authorisation, or suppressed to the point of impeding access to the same oceanic benefits to other international (present or future) actors (including the individual). In short, a given activity cannot empty out another potential activity when it occurs in the Area.

Regarding this, Freestone (2018, pp. 135–137), in terms of biodiversity, and Scovazzi (2000, pp. 213–215), in terms of genetic resources, report the difficulty in managing the Area in this sense, when it comes to mining.

According to experiments conducted in the Pacific Ocean (DISCOL project), close to Peru, in the so-called Peru Basin, a polygonal in the Area was the object of a simulation of possible exploitation of polymetallic nodules (just like the requests, still in the exploratory phase, prevalent in the CCFZ). The damage simulation took place in 1989 and, although tests and the protected isolation of the polygonal seabed itself were isolated from relevant anthropic interference, the recovery of the biotic and abiotic elements degraded by the simulation was not observed (Vonnahme et al. 2020, p. 1).

Notice, however, that ISA, although it has the power to authorise the exploitation of natural resources in the Area, will only do so in relation to non-living resources (Freestone 2018, pp. 135–136).

Hence the importance of this principle for this work: it is not possible to mine funds, due to the current state-of-the-art, without permanent and indelible (*i.e.* non-recoverable) consumption of the biotic and abiotic portions of the mined area. How, then, to legally combine this idea of being authorised only for the mineral exploitation (mining) of the Area and, at the same time, access to the associated biotic elements?

Both Freestone (2018, pp. 135–136) and Scovazzi (2000, pp. 213–219) agree that UNCLOS III regulated situations, even if associated and/or in an integrative dialogue with other regimes (article 304, UNCLOS III), as in the case of the Convention on Biological Diversity (CBD) 1992, with its protocols in Nagoya and Cartagena, respectively, regulating these portions of the Area's ecosystems. However, these inter-regional dialogues provoke doubts and obligatory escapes that forced both authors to agree, again, about the doubts or the needy protection of the marine environment, which suffers anthropic impacts unprecedented in the history of the relationship of man with the sea.

The principle in question is a need for an international and binding standard (Res. 2749, of the XXV UNGA), in view of the issue of the prohibition of appropriation of the funds: among the powers of possession or property, that of excluding the other (sequel) justifies it as an effective right. The owner of property/property (as a means of production) is the sovereignty regime outlined in 1648 by the Treaties of Munster and Osnabruck, one of its most evident characteristics being the power to exclude the other (State and its subjects) from acting in certain territory or goods (Ferreira Filho 2012, pp. 88–89).

As such, excluding multiple uses from other States due to the exhaustion of possibilities by a single category of use or, for the same reason, denying humanity (in its broadest sense, including future generations) goes against the very reason this polygonal policy exists.

Later, when dealing with sustainability in the Area, we will understand that one of the forms of appropriation is, precisely, the irreversible degradation and pollution of a biome (Tharpes 1989, pp. 606–610).

For the purposes of UNCLOS III (article 1 (4)), pollution is defined as follows:

> Pollution of the marine environment "means the introduction by man, directly or indirectly, of substances or energy into the marine environment, including estuaries, whenever it causes or may cause harmful effects, such as damage to living resources and marine life, risks to human health, obstacles to maritime activities, including fishing and other legitimate uses of the sea, changes in the quality of sea water, with regard to its use, and deterioration of recreational sites".

Therefore, although it was not, at first, formally called or considered in this way, the principle of harmonisation of activities in the Area and the maritime environment had been recognised by the doctrine as a result of articles 1 (4), 2, 89, 117, 113 to 137, and 147, UNCLOS III (Scovazzi 2000, pp. 206–219; Goettsche-Wanli 2018, pp. 46–56), although its recognition as a norm international law applicable (and not just a precept) to the Law of the Sea regimes has been derived from Case 17 ITLOS (ITLOS 2011, p. 28).

For the principle not to be a contradiction in terms, and to be betrayed in meaning and application, this principle must be understood by looking at its exercise in time. In this sense, Marotta Rangel (2014, pp. 466–468 and 480–483) enunciates about the Area and its multiple uses, pointing out that it is necessary to manage not only these uses and accesses, but also the extraction of wealth from it, going to detail the difference between resources (article 133, UNCLOS III, which points out all the abiotic elements of the Area system that can be valued) and ore (which is nothing more than the extracted resource), working on the idea of a central agency (ISA) that has the interests of the International Society in relation to the Area over time.

It is this temporal dimension of law, instilled in the thought of international law since the 1980s, of the Section XX, both at the precept level and at the level of the hermeneutic elements of production of the legal norm, notably represented by the "Our Common Future Report", that the principle of harmonisation of activities in the Area and in the maritime environment finds ways to solve one of the questions in this work, such as will be seen in Chap. 3 of this book.

Participation of Developing States in the Activities of the Area

This principle, according to Menezes (2015, p. 156), aims to allow the

> Effective participation of developing States in activities in the area must be promoted considering their special interests and needs, whether those that have no coastline, whether they have it, but are in a disadvantaged geographical situation. Activities in the area will be shared among all from the perspective of a democratic space, but essentially of collective solidarity for all peoples.

The syntax of this principle of the Area with the general principle of the common but differentiated responsibilities (CBDR) is evident in that not all UNCLOS III

States Parties have the same conditions of access to the sea (some do not even have the coast, like Paraguay), although to all access and benefit-sharing is guaranteed (UNCLOS III, article 148).

To deal, then, with this principle, which is revealed (later) as a specification and consequence of the CBDR, we take a step back to understand this controversial principle of international environmental law established in the Declaration of Rio/1992.

The proposed analysis goes through a simple understanding: how can CBDR, a principle enunciated in Rio/92, be binding as an object of analysis of international law applicable to the subject? In fact, the CBDR's isolated analysis forms the conclusion that the principle is a soft *law*. However, one cannot forget two variables: (a) this principle carries with it an expected current behaviour; that (b) when submitted to the expressions of Parts XI and XII, Montego Bay/82, brings a new identity to the expected behaviour of the States Parties. It is worth saying: notions and environmental principles, although soft law, are fully applicable when assumed by unilateral declaration (Schrijver 2008, pp. 238, 254–255 and 347–348; Scovazzi 2000, p. 103), as well as the state-of-the-art of scientific and social development are expected and imposed hard law behaviour (ITLOS 2011, pp. 46–51). The State is not obliged to a soft law, but a hard law that incorporates it into another norm as content (*i.e.* the "paradigm" informing the hermeneutic process that forms the norm) of observance.

Added to the categorisation of the area as a *commons* of the International Society, by virtue of Res. 2749, of the XXV UNGA declared as common heritage of mankind (CHM) right in the first item of the Resolution.

In view of this phenomenon, nothing is more appropriate than considering the participation of developing States in Area activities a principle of international law, whose compliance and effectiveness obeys an optimisation criterion peculiar to the realities of the assets it seeks to protect.

However, although the form of manifestation and legal linkage of the CHM is disparate between protected objects, there is still a minimum content of work which identifies its occurrence and influence on the behaviours of the actors. Wolfrum (2012, pp. 4–5) started from an analysis centred on the CHM principle in the Area, in comparison with other occurrences of the same principle, and stated minimum parameters for the recognition and application of CHM. See:

(a) A regime for the use of the patrimony object of CHM, which will presuppose international cooperation for the safeguarding of the patrimony, the international administration for the patrimony and the regulated/regulated use
(b) The sharing of benefits
(c) The dimension and the intertemporal value of the patrimony
(d) The peaceful use of heritage

Wolfrum (2012, pp. 4–5), applying the concept to the species, a judge of ITLOS at the time of the judgment of Case 17, outlined these *Commons* called seabed and described the obligations inherent in the Case Report and whether CBDR applies to them or not.

Hence, it is perfectly justified for ITLOS to claim the declarations of the States Parties (albeit in non-binding diplomas), as well as the state of development of the dialectic between the actors of international law, to apply CBDR as a rule of the Law of the Sea, in full judgment an important aspect of the Montego Bay/82 regime. And we go further: to the extent that this principle stated in *soft law* composes a court decision given by hard law (binding), two effects emerge.

First, the principle becomes binding because it serves as a precedent for ITLOS itself (ICJ Statute, article 38).

Second: the principle is recognised as a material component of compliance and effectiveness of the treaty itself, composing the content (or, in the expression of the general theory, "ethical paradigm") of the legal norm itself by hermeneutics of the precept.

Given the possibility of implementing CBDR in the Area, it remains to be seen how this principle would be applied in this region.

It is important to note that Costa de Souza (2019, pp. 113–119) understood that the application of the CHM in Case 17 ITLOS was so peculiar that it even de-characterised the claimed principle, reaching the limit of conceptually approaching the common heritage of mankind.

In the judgment of Case 17 ITLOS, the Court made material assumptions for the application of the CBDR in accountability of States Parties to mining events. See ITLOS (2011), which provides:

1. All states, regardless of their social, economic, or (geo)political conditions, are entitled to sponsor entities for mining in the area, subject to due procedure before the Authority (ISA).
2. All states have agreed that the signatory parties have different social, economic, or geopolitical conditions and, as such, are recognised when dividing areas to be explored and extractive and financial exploitation regimes (especially in articles 148, 150 (c) (g) (h) (i), and 153, Montego Bay/1982).
3. The environmental occurrences in the Area and their responsibility by the Sponsoring State are similar, when observed from the point of view of risk, with the vessel registration regime, in the international Law of the Sea, regulated by Montego Bay/82 (Costa de Souza 2019, pp. 116–119).

For ITLOS, observing the case in question, the CBDR (and, consequently, access to nations at different levels of development) in the field of seabed mining operated by differentiating the rights assumed in an international legal norm with a view to the solution the global problem of inequality in the possibility of accessing and obtaining mineral wealth. In this case, CBDR acted regarding the unequal redistribution of areas (reserved areas), the operation of the company or by the State-Sponsor regime, in what Egede declared to be the Nauru-Tonga system (explained above), in a specific application to those we observed in Honkonen (2009, p. 257), but not to the criteria for checking accountability parameters, their extent or obligations associated with their taxable event.

It is not said, in turn, that Montego Bay/82 imposes technology transfer for the area's mining issue. Since 1994, this technology transfer has ceased to be a mandatory subject (cogent) of public international law and has been transferred to the private international law regime of contracts between States Parties and sponsored companies.

Nor is there any talk, due to the similarity of State-Sponsor systems for mining with the flag system for navigation, on different responsibilities of a State-Sponsor (and the *liabilities'* consequence) regarding their duties towards the environment of the Area and marine surroundings affected by mining.

Therefore, initially, we consider the idea that CBDR is not applicable in the Area to be inaccurate. Indeed, it is, but within the scope of the distribution of rights and resources among States Parties.

Meanwhile, regarding the recognised environmental responsibility in the Area (Part XII and article 148, Montego Bay/82), CBDR is normally considered not applicable.

However, another point of view can be observed: consider that the CBDR exists in the dynamics of rights and obligations of the Area, due to the discipline of reserved areas and division of results (Costa de Souza 2019, pp. 116–119). Thus, the principle in question affects and is relevant to the discipline of the Area. What is important to observe, from this other point of view, is the fact that CBDR is a principle and, as such, an optimisation mechanism based on the material and/or infrastructural measure of possibilities, including political ones.

Seeing that the principle is the norm of compliance by a differentiated criterion: it is necessary to comply as much as possible by an "optimal" criterion estimated according to infrastructural (or material) elements (Wolfrum 2018, p. 103).

In Case 17 ITLOS, judges have no previous seabed mining experience, so they used other material elements in analogy. Let us see: the object of the consultation was the responsibility of a Sponsoring State for environmental damage.

In addition to seabed mining, another system had the potential to generate significant environmental damage and implied control and grant regimes by the State Party and the private sector: the registration of merchant vessels, also called "flag concession".

It is observed that in this regime of concession of flags, accidents with environmental damages (Lawand and Valdez Silva 2019, pp. 3–5) have a peculiarity that, equally, did not go unnoticed by the judges of Case 17 ITLOS: the duty of care and minimum standards for checking and quality in the duty to prevent damage and recover it properly.

The dismal experience of the flags of convenience in a sea regime in force and its intrinsic relationship with the occurrence of major environmental accidents, in addition to Res. 2749 (XXV), UNGA, making the Area a common and universal domain, imposed a reinterpretation of the CBDR principle and "No Harm" (Schrijver 2008, pp. 303–305) in the regimes for the use of maritime spaces: CBDR was not eliminated from the Area regime, but the optimisation of its implementation for the exclusive purposes of accountability of the Member States part was neutralised with the purpose of not having nations with "Convenience" sponsorship for the mining of the area, risking universally considered patrimony and nullifying the effectiveness of article 191, Montego Bay/82. Thus, whoever has the initiative to propose a mining process before ISA, it will have to guarantee itself with the means it considers possible in the case of accountability: the CBDR incidence phase comprises the choice of the area (and, when operational, revenue sharing). Once the process has started, the mandatory paradigms (including liability for environmental damage) are formal.

Hence, the result of the case in question in stating that the Sponsoring States' liability derives from: (a) the inability to comply, whether commissive or omissive, in the duty to establish legislation and a duty of protective diligence to the Area, as CHM; and (b) the occurrence of the damage or its imminence (Wolfrum 2018, pp. 103–104).

Thus, the CBDR takes shape in Montego Bay/82 for the fundamental design of Case 17 ITLOS, which combines it with two other principles: that of CHM, incident, and assumption of activities in the area and that of *No Harm*. It is possible to understand that the proposal of the mining activity and its (business) results have very identifiable contours as CBDR, but their incidence, even in the whole activity, falls back on the transfer of technology and the duty to restore the degraded area to the status quo ante, the view of the order of non-aggression to the patrimony of the other international actor and of the common responsibility in relation to this patrimony.

Therefore, with respect to mining in the Area:

> It must be a space that allows the affirmation of all peoples and whose resources are destined to the States that need them most and have structural limitations to invest in activities in it (Menezes 2015, p. 156).

What Is Conclusively Understood About CHM and Principles?

What is kept of the CHM as an Area regime, in short, for the purpose of this work, has its centre in the following assertions:

(i) It is a principle that has a reason for being in different permitted uses among actors that share it, and these uses are limited by the various simultaneous, future, and non-prohibited categories of access between these actors, these uses being directed in lanes or logical propositions seen above.
(ii) The 1994 reform of UNCLOS III and Part XI may have weakened this CHM in the concept originally thought by Pardo. However, this weakening, currently, occurred only at the level of effectiveness and did not disappear or denature the CHM in the Area, nor its limitations of inherent multiple uses.
(iii) A single use that renders the Area useless for other uses, even if nominally prescribed, becomes illegal and illegitimate, as it constitutes appropriation, in view of the other rules of the Area's legal regime.

2.1.2 The Area as a Global Public Good (GPG)

In 2008, at an event in New York City, a member of the management of the hedge fund FrontPoint Partners LLC, Steven Eisman, discussed the health of the stock market and the securities market (*securities*) in the United States with Bruce Miller, a senior executive and shareholder of Bear Stearns Bank.

Eisman stressed that the problem was not (and never was) the lack of money or liquidity. But, yes, an absolute and complete lack of governance (which even tended to the barriers of language itself) that would cause the American financial market to collapse (which happened that same month and, in fact, was striding occurring in the same minutes in which he was speaking).

Both public and private systems of supervision and control of this market, intricate and interdependent, have failed to observe the obvious: that the strength of a chain is measured by the weakest link. In financial terms, the strength of an IF is measured by the solvency of its minority bond rating credit. And the failure was not in the lack of data: data existed, and Eisman was able to access them. The failure was, therefore, in their public treatment systems.

But what does this have to do with the Area?

Both the Area and the globalised financial market are common goods, shared by all, that influence and keep private and national objectives and plans, but affected, influenced by private or national (non-global) decisions.

The Area is, by express and straightforward convention, of a Res. 2,749 - 22a., It is qualified as a CHM, and its legal characteristics, starting with the legal nature, can be observed under another point of view in international law: the "global public goods" (GPGs), just like the financial market (whose biggest node is the New York Stock Exchange).

To exercise this new point of view, let us look again at the comparison of the ocean with the legal figure of the worldwide used compassion (Grotius 2002, p. 87): private and national needs have, over the years, passed through the realisation of a form of common use, guarded by common values and objectives. Applying the concept to the species, it was the free and peaceful use of the seas that provided the nations of the world with their private and national developments according to their own plans.

The compassionate and the seas are public goods, this being global, according to economic theory, because (Kaul et al. 1999, pp. 3–4):

(a) Non-excluding: which means that the use of the good by one individual does not prevent the use by another; and
(b) Non-rival: which means that the use of the good by one individual does not diminish the quality and quantity of use by another

The seas and in particular, the international seabed, in this case, have historically been a common use asset, allowing all international actors (members or not of the International Society) to use it as raw material, without denaturing it. lo: the private objective was fulfilled with the rational use of the public good. By UNGA act, establishing a legal regime specific to the international seabed, in which its uses were multiple and subject to a trans-generational equity related to sustainable development. It means to say that one generation cannot benefit from the area, rivalling another or between peoples with itself.

Whether due to custom in the Law of the Sea, or due to the binding force of the UNGA act, the seabed fulfils the double economic requirement to qualify it as a GPG.

A reverse thinking would be more illustrative: a fact, a particular/national action or omission that would put this global and common public good at risk would generate "externalities" (risks to your activities/plans/private/national projects) that are also national/private to other international actors. Those affected, in turn, would behave according to their new, poorer reality in the dialectic between international actors, affecting and (even) exporting this externality to the particular "offender", causing the common well-being and the GPG to enter a deleterious vicious cycle to all parties and to the GPG itself.

At this point, a less institutionalist look (which gave rise to the CHM) brings us to a more recent concept of international law, that of "global public goods" (GPGs).

GPGs are things, institutions, principles, or global values that are useful or necessary to international actors, collectively or individually considered, for the fulfilment of their interests (national or private) and whose malpractice risks, directly or indirectly, the development of all actors.

This concept is not new and summarises the understanding of Nye Jr. (2002, pp. 239–243) and Kaul (2000, p. 23) about this different approach to the theme.

Unlike CHM (which does not allow sovereignty and, consequently, appropriation), GPG can be public or private, it can be subjected to a public or private regime, but it has peculiar and common characteristics:

(a) Can be subject to private actions and objectives that have a significant impact on everyone who interrelates with the GPGs
(b) That the externalities resulting from the management of GPGs are suffered by several actors of the International Society (public, private, organisational, or individual)

This approach to global issues through the concept of GPGs carries some forms of recognition and action that Kaul et al. (1999, p. 10) summarise well:

(1) An action that involves the best chance of success: in the face of a crisis caused by the malpractice of this GPG it demands an effort called "best-shot", *i.e.* an action by the global actor of excellence in solving or preventing the problem. In fact, from the point of view of private action for communal benefit, the very resolving result promoted by this actor will be a GPG.

    As an example, Kaul et al. (1999, p. 10) point to a centre of excellence (even if private) as responsible for developing a clinical form of control of a disease of worldwide interest.

    In our case, we could point out the efforts of MIT in creating methodologies for the management and regulation of categories of degradation caused by mining (of polymetallic nodules) in the Area (Freidah 2019), that is, the plume of particulates that it navigates. The water column harms the Area's macro- and mesofauna and the immediately adjacent water columns.

(2) A joint action of cardinal actors involved in the problem: as an example, Kaul et al. (1999, p. 10) points out the actions of preservation of forest systems among actors who share them in greater proportion and/or have greater problems of deforestation. Both the action itself and its results, as well as the forest services that these environmental assets offer, are GPGs.

 In our case, we can point out the sponsoring nations of the Clarion-Clipperton Fracture Zone (CCFZ), in the joint efforts of the draft Regulation 45, ISA, to identify the effects of possible degradation caused by mining in the Area or the call of stakeholders on mining in the Area to Pretoria/2019 to find sustainable solutions that combine exploitation with article 194 and 145, UNCLOS III.

(3) A global sum of efforts, in which all countries would be compelled to participate.

(4) The "weakest link", to say that addressing issues related to GPGs and its management depend, precisely, on doing so at its most critical point, at the weakest link in its chain of events.

Kaul (2000, p. 22) points to the example that eradicating a disease on the planet depends on efforts with countries that are more sensitive to this disease and/or less able to manage it.

At this point, for the purpose of analysing mining activities in the Area, we switch our focus from Eisman to Egede. The mining regime established in Part XI states that any State Party may, by means of its own procedure, request the exploration, and develop the exploitation of international seabed for the purpose of mining.

Egede (2018, pp. 157–184) works on this issue from the point of view of the environmental obligations of articles 145 and 196, UNCLOS III, in the light of developing States and the experiences (of UNCLOS III itself) about flags of convenience in navigation (BDC). In other words: Egede recognised the Area's environmental issue as a GPG, even though it did not use this expression, and treated it by its likely risk bias.

This is the sensitive point when we look at the Area and its mining as GPGs: the fulfilment of these obligations by the sensitive actors considering the past and analogous experience.

Dealing with environmental responsibilities in the Area for mining purposes takes us to Case 17 ITLOS, the content of which, for the purposes of this analysis in the light of the GPGs, can be summarised in:

(a) Recognises the existence of the reserved areas in the Area and the areas in which ISA, through exercise of preference through the company or another entity, will promote mining under the most favoured/preferential treatment regime to developing States Parties (UNCLOS III, article 154).

(b) Except for the principle of common but differentiated responsibilities (CBDR), it recognises that this differentiated treatment does not apply to issues relating to mandatory environmental responsibilities and to debts arising from environmental damage inherent to States Parties that sponsor mining activities in the area, neither do they apply to the requirement of applying the best environmental practices (in the mining production process).

(c) All the principles contained in the Declaration of Rio/92 are applicable and considering the exploratory phase of mining activities and articles 146 and 195, UNCLOS III, Principle 15 - Precaution, of said Declaration.

However, Egede (2018, p. 164) points out that Principle 15, of the Rio Declaration/92, has as its application criterion the application by the States of "according to their capacities", which, in theory, would open clear for the broader participation of developing States.

However, the approach of the ecosystems of the Area in the light of the GPGs by the "weak link" criterion implies asking who the "Developing States" are for the purposes of UNCLOS III, article 148?

It implies more: if the same window that allows for broader participation, also allows the creation of "sponsorship of convenience" figures, in analogy to the BDC, then what compliance measures should ISA and international actors adopt?

The decision about what is "Developing State," mentioned and not defined in article 154, UNCLOS III, although recommended by ISA when issuing act ISA/22/A/CRP, article 3 (2), has not yet been taken and will necessarily depend on a political choice by some technical analysis criterion (Egede 2018, p. 164).

What is closest to what I need to point out is that this development criterion will be linked to variables such as GDP, HDI, and other indicators economically linked to the country's capacity.

Part of the mandate that the Area is of common benefit, as BGP is, but its practical mechanisms of access occur due to the degree of technological development of the State in reference. In this light, while this access is marked by information and technology (Stiglitz 1999, pp. 308–325), this information and technology becomes BGP itself when: (a) areas on the international seabed of access to mining are reserved. "In Development" and the transfer of technology, even though after the 1994 reform of UNCLOS III is not mandatory, it is undoubtedly an issue of value and subject sufficiently important to form a precept in the Convention (UNCLOS III, article 5, Annex III); (b) countries that, by criteria of analysis, others are considered "in Development".

In this way, the process of conferring the mining rights of the Area, as GPGs that is, there must be actions directed at the "weak link" of the global production and distribution chain of this right recognised in Part XI, UNCLOS III.

Therefore, from the mining concession process on screen, already seen before, two phases of control stand out.

First group of requirements (UNCLOS III, article 3 and article 4, Annex III) (or first check): checking of attributes of the pleading State Party to comply with the technical requirements for direct exploration in the portion of the Area pleaded.

This first group of requirements was pointed out by Case 17 ITLOS, when it commented on article 139, UNCLOS III, as the scope of due *diligence* to be suffered by the State Party pleading for the right to exploit (with a view to exploiting), regarding its domestic legislation and the capacity to apply it. [6]

On this point, two considerations: the first, ISA itself, by analogy and by technical and direct criteria, pointed out which are the nations that need material subsidies for the exercise of mining in the Area. In this way, and at least for the Law of the Sea regime, we have reference and regulation for article 148, UNCLOS III, the view of the rules of interpretation and integration of the international treaties by the Vienna Convention-I/1969, Section 3.

The issue of article 148, UNCLOS III, we have the second: how UNCLOS III gave access to the Area to the nations included in article 148?

UNCLOS III, in its Part XI and Annex III, gave two alternatives to the direct exercise by the State Party of mining activities in the Area, namely: the first, the regime instituted by the Area reserve and the company, under the terms of article 9, Annex III, UNCLOS III. However, until the closing of this text, the company was not yet installed (although founded) in the ISA/UN system.

The second is the Sponsor State regime. And this is where the second check takes place, to prevent the "weak link" in the chain from breaking, harming GPGs and its dependents (*i.e.* the International Society and other actors).

Regarding the system used by the Sponsor State, we have the provision of article 4 (1 to 4), Annex III, UNCLOS III, interpreted and whose responsibilities (in environmental matters) were described in Case 17 ITLOS. It is in this trinity of support (the production of the international legal standard, its interpretation by the authorised actor, and its interpretation by the competent judicial body) that the duties of the Sponsoring State and the entity (sponsored) in the mining process of the Area are structured.

Therefore, those duties inherent to the State Party do not change (not least because the perspective is the Area's environment as a good at the service of all), but will be performed by another actor, the entity. Therefore, finally and as a technical alternative, the entity will supply the deficiency of technical parameters when the State cannot reach it.

The first problem lies in the requirement that the entity, to be recognised as valid and able to take the exploration contracts for the Area and execute them, it must comply with the following requirements (UNCLOS III, article 4, Annex III):

(a) The dictates of article 135 (b), UNCLOS II; and
(b) Be national or under (indisputable) corporate control of the Sponsoring State

---

[6] *UNCLOS III*. Annex III. Article 4: "4. The sponsoring State or States shall, pursuant to article 139, have the responsibility to ensure, within their legal systems, that a contractor so sponsored shall carry out activities in the Area in conformity with the terms of its contract and its obligations under this Convention. A sponsoring State shall not, however, be liable for damage caused by any failure of a contractor sponsored by it to comply with its obligations if that State Party has adopted laws and regulations and taken administrative measures which are, within the framework of its legal system, reasonably appropriate for securing compliance by persons under its jurisdiction. 5. The procedures for assessing the qualifications of States Parties which are applicants shall consider their character as States."

Here, Egede (2018, pp. 178–182) sees the first breaking point of the “weak link”: the recognition of nationality of individuals or legal entities and the recognition of the existence of these are matters that are subject to national legislation (Aust 2010, pp. 11–12; Abi-Saab 1996, pp. 79–82 and 126–127). The statement finds support and echo in the *Barcelona Traction*/1970 case.

It is in national legislation that the first temptation for the “international environmental fiscal war” resides:[7] the “race to the bottom” between States Parties in the quest to become Sponsor States can facilitate the structuring of (sponsored) entities that, in normal situations, would not meet the governance, transparency, and compliance requirements necessary for mining activity in the Area.

Two mechanisms discourage this corporate practice at the international level.

The first resides in the simulation and/or fraud institute of article 39 of the Vienna-I/1969 Convention. In this case, the fraud recognition mechanism is in the objective elements of the link between the entity's legal entity and its controllers with the Sponsoring State (Lawand and Valdez Silva 2019, pp. 1–12), in view of the unambiguous interpretation of ITLOS in Case 17 ITLOS for article 139, UNCLOS III: the perspective is taken from the risk to measure obligations. Any application of CBDR is excluded, with a norm of equal application to developed and developing countries prevailing.

The second element of discouraging the use of inconvenient corporate types is the accountability regime explained by Case 17 ITLOS: the deficit legal structure promoted by the Sponsoring State to its sponsor imposes in direct assumption of all the environmental responsibilities inherent to the mining activity required in the Area, under the terms of the contract (Case 17 ITLOS).

But an eventual lack of transparency and governance requirements at the business level would be sufficient ballast for the imposition of articles 139 and 153, UNCLOS III, making the Sponsor State fully responsible for the occurrences and environmental obligations of mining activity in the Area?

The answer is in the affirmative. Let us remember that the same Case 17 ITLOS imposes the Declaration of Rio/92 as an application element for understanding and safeguarding the environment, when so required by UNCLOS III. Although there is a prestige to Principle 15 (Precaution) in the manifestation of the specialised Chamber of ITLOS, all the others are also applicable. These are reasons that lead us to observe so much: first, there is an implicit or express mention of other principles contained in the same Declaration: now as an example of application, such as Principle 13 - polluter pays (ITLOS 2011, p. 71), now except for the incidence of a certain principle, such as o Principle 7 - CBDR (ITLOS 2011, p. 48).

Thus, there is no reason for us not to understand in the list of environmental responsibilities of Sponsor States and other actors in the mining activity in the Area, the full compliance with the duty of information and transparency transcribed in Principles 10 and 19 of the Declaration of Rio/92.

[7] In reference to the ICMS tax wars between units of the Brazilian federation, see Dulci (2018).

In fact, this duty is regulated by Annex III, UNCLOS III (article 4 (4)) and ISBA/7/LTC/1/Rev.1/2002: the precept keeps the confidentiality of certain sensitive information, whether for reasons of security of a given State, whether due to issues related to intellectual property and intangible assets (which, due to the 1994 reform of UNCLOS III, are no longer obligatorily shared).

So much so that the contract between the Sponsoring State and the sponsored contractor, according to Annex III, UNCLOS III, has no opening order: it is a national and sensitive object. However, this same contract cannot, according to Case 17 ITLOS (2011, p. 65) serve as a parameter of exclusion of responsibility of the Sponsoring State for the environmental regulation contained in its object. The environmental regulation to which the contractor is subject must be public, and, therefore, included in the sponsor's positive law. Only the deposit of the contracting certificate with ISA is required from the sponsor, who immediately assumes his responsibilities under discussion.

Therefore, if there are precepts to transparency (*i.e.* confidentiality), it means to say, by logical process, that it is the rule.

Second, there is an express mention of the application of the Declaration, with no exceptions other than the specific ones (Wolfrum 2018, pp. 103–104).

Let us note that Kaul (2000, pp. 22–23) had already pointed out the fact that GPGs would be very much subject to the unilateral acts (which the author calls "private") of management rather than to collective actions. This is another significant indicator of the occurrence of the concept of GPGs and its incidence in the Area and its mining process, giving them this legal nature.

Let us return to the normative disincentives to the "Sponsor of Convenience". The second is the duty of diligence and the creation of an environmental legal framework that guarantees the mining operation (in its exploratory and exploitation phases), the implementation of the precautionary principle through rules that establish the Environmental Impact Study (EIA).

Thus, ITLOS (2011, pp. 70–71) recognises:

> Sponsoring States have two types of obligations under the Convention and related instruments:
>
> A. The obligation to ensure that sponsored contractors comply with the terms of the contract and the obligations set out in the Convention and related instruments.
>
> This is a "due diligence" obligation. The sponsoring State is obliged to make every possible effort to ensure compliance by the sponsored contractors.
>
> The standard of due diligence can vary over time and depends on the level of risk and the activities involved.
>
> This "due diligence" obligation requires the sponsoring State to act within its legal system. These measures must consist of laws and regulations and administrative measures. The applicable standard is that the measures must be "reasonably adequate."
>
> B. Direct obligations that the sponsoring States must fulfil regardless of their obligation to guarantee a certain conduct on the part of the sponsored contractors.
>
> The fulfilment of these obligations can also be seen as a relevant factor in the fulfilment of the sponsoring State's due diligence obligation.
>
> The most important direct obligations of the sponsoring State are:
>
> (a) the obligation to assist the Authority established in article 153, paragraph 4, of the Convention.

(b) the obligation to apply a precautionary approach as reflected in Principle 15 of the Rio Declaration and set out in the Node and Sulphide Regulations; this obligation should also be considered an integral part of the sponsoring State's "due diligence" obligation and applicable beyond the scope of the two Regulations.

(c) the obligation to apply the "best environmental practices" established in the Sulphide Regulations, but also applicable in the context of the Node Regulations.

d) The obligation to adopt measures to ensure the provision of guarantees in the event of an emergency order from the Authority for the protection of the marine environment; and (e) the obligation to provide compensation resources.

At this point, Egede (2018, pp. 160–164) enunciates yet another weakness of this "weak link" in development, namely: in the same logic as the internal legal forms of constituting national juridical persons of the Sponsor States to constitute a field for a new "Race to the bottom", due to the loosening of requirements.

Krisch (2014, p. 4), because of this risk to GPGs (in gender), the sovereignty system based on the Peace of Westphalia/1648, either by the tendency of inaction of the actors responsible for the achievement of the common objective, or by the rationalisation of the cost- opportunity for the prisoner's dilemma, the sovereign consent variable undermines the administration of a GPG.

Three measures are used by the regime in question for the purpose of reducing the risk of a convenience sponsor who uses inadequate internal regulatory criteria that could risk the environment of the international seabed.

The first is the criterion derived from UNCLOS III itself, from the state-of-the-art regarding environmental normative criteria of prior due diligence and previous environmental studies, contained in article 153 (4) and (5), UNCLOS III.

The second comes from what Egede (2018, pp. 165–171) called the "Nauru-Tonga system". These are developing Polynesian States Parties (Nauru, Tonga, Fiji, Tuvalu), which saw in the mining area an opportunity to develop economically in another segment. As "weak links" (in Kaul's terminology), they opted for the regime of Sponsor States and, in turn, had to adapt their internal regulatory frameworks to comply with the dictates of article 153, UNCLOS III.

As a way of not starting the fiscal war between these more fragile countries and transforming the sponsorship into a sponsorship of convenience, these States have joined together in an assistance office for this purpose Egede (2018, pp. 165–171), which, in turn, in turn, it advised them to standardise the environmental legislation that would give rise to the sponsorship of entities in seabed mining by any of these countries.

Within the scope of this advisory, Nauru and Tonga provoked the ISA Legal and Technical Commission about the environmental responsibilities of the Sponsor States, which was shown in document ISBA/16/C/6, proposed to the Authority Council, which, in turn, questioned ITLOS (Case 17 ITLOS), through the deliberation of the ISBA/16/C/13 Council.

Therefore, and being characterised the ecosystems of the Area as GPGs, in a timely chapter we will see how this legal nature talks with the requirement and the current impossibility of RDA in the mining of the Area.

2.1.3 *Dialogue with Other International Regimes and Precepts and the Sustainability of the Oceans*

It comes to the critical point of the text in which it is mandatory to understand how sustainability operates (and this principle as key stone of Law of the Sea regimen with other international regimes) for the international seabed. We will analyse, by virtue of Case 17 ITLOS and its links to the CHM and sustainability built in the Declaration of Rio/1992, how this second influencing principle of the Mining Code works.

After this phase, we will have enough arsenal to address the administrative mining regime imposed by ISA and the question of foundation: according to the international standards that regulate the international seabed and its utilities, the activities are guided by a principle of sustainability and CHM and, therefore, if there is an activity that effectively degrades the area's environment, is there an RDA obligation to the executor/sponsor of the activity?

In the words of Boyle (1995, p. 174), "[s]ustainable development 'has become the central objective of international development policy (. . .)". Despite the illustrative statement, Boyle himself recognises that it does not explain much about what sustainable development is. Therefore, an explanation of this theme has been chosen according to international environmental doctrine and the linkages affixed in Case 17 ITLOS.

Boyle (1995, p. 174) points to this origin as coming from the publication of the United Nations, in 1980, of the document World Conservation Strategy. However, Dallari Bucci (2001, pp. 56–59) points out that the concept of sustainable development was brought up by the Brundtland Commission, instituted by the United Nations General Assembly, whose work began in 1984 and ended in 1987, when the publication of the "Our Common Future" report.

This commission arose from the decisions taken at the United Nations General Assembly, in accordance with the deliberations of the Stockholm Conference, in 1972, and produced the document on which discussions took place during the analogous Conference in Rio de Janeiro, in 1992.

We can trace the first lines of what sustainable development is, based on the idea of a compromise between economic growth and the environment, which does not necessarily mean zero growth, points out Boyle (1995, p. 174). The concept of sustainable development was taken from a Report (UN 1987, p. 16):

> Humanity has the ability to make development sustainable to ensure that it meets the needs of the present without compromising the ability of future generations to meet their own needs. The concept of sustainable development does imply limits - not absolute limits but limitations imposed by the present state of technology and social organization on environmental resources and by the ability of the biosphere to absorb the effects of human activities.

From this concept, we can already verify, even at first glance, that it is a developmental process. The balance of production relations with the environment (both social, as will be shown below) does not find a static optimisation, but it develops in a dynamic process according to choices.

These choices are based, among other variables, on the establishment of the needs a given society is required to meet. These needs, which influence development, will be computed in a global development strategy, in which, the major preponderant factor in prioritising these needs will be said by the second part of the concept of sustainable development.

Therefore, sustainable development depends on decisions (options), necessarily political, although necessarily informed by scientific language (and its limitations, as in the concept).

In this way, all the options taken in the composition of this global strategy referring to sustainable development, whether social, economic, legal, military, scientific, or any other, will be (or would have to be), invariably, political. And if policies, they will seek their implementation through their own language.

Forged by the Rio de Janeiro Declaration of 1992 (with rudiments of the Stockholm Declaration of 1972), it is a point of convergence of international diplomas, among them UNCLOS III, at the international level, sustainable development, dedicated to the protection of quality of life, for through the reformulation of the relationship between man and the environment, it is raised to the level of fundamental right, uniting three dimensions of conceptual (Schrijver 2008, p. 288): preservation/conservation, development, and resources equal or greater to the future.

If Schrijver (2008, p. 288) points out that sustainable development is a principle recognised by international treaties, forged over 25 years and still under development, Carlarne (2014, pp. 9–13) points out the brutal difficulty in advancing efficacy of the same principle or its recognition (especially when worked against ancient principle-logical grounds such as sovereignty, based mainly on the Peace of Westphalia/1648). This author gives reason, among others, to a wrongly avowed "lack" of gründnorm that supports sustainability (as it happens with the sovereignty itself).

Both authors affirm the precept in public international law. But each one does it in a different way.

Schrijver (2008, p. 289) contradicts Carlarne and points out that the principle of sustainable development is indeed "grounded": through the "Our Common Future Report"/1985 (or Brundtland Report, as it is well known), commissioned by UNGA, and the Declaration of Rio de Janeiro/1992.

The problem: both the report and the 1992 Declaration are qualitatively, and in isolation, non-binding.

The solution to the apparent doctrinal-normative impasse comes from reading Marffy-Mantuano (2011) about the same principle in the light and as an interface of two international regimes (the environmental and the sea) for marine biodiversity.

The first solution, which Marffy-Mantuano (2011) calls constrictor, derives from UNCLOS III, a binding rule, as the framework of a whole regime of international law, namely, the sea regime.

The author, taking this coded source, points to marine biodiversity as an object of commitment linked to sustainable development by express preceptory determinations (ranging from the preamble to specific levels of provisions in terms of fisheries and the marine environment). Her interpretation makes sustainability a mandatory precept.

Considering that ITLOS, when legally assessed, is a mandatory and binding means of assumption of controversies, the reservation of which is impossible to be made upon the adoption (signature) of the Convention by any of its state's parties. Therefore, in the judge's work in building the legal standard based on the precept, let us analyse ITLOS from the point of view of sustainability.

Case 17 ITLOS imposes, as a requirement for the accountability of the explorer and/or exploiter of resources in the Area, a prior due diligence, and police (administrative) duty to the Sponsoring State of the activities. This duty of care is nothing more than the sponsor's inspection of the sponsor's activities (on a preliminary and operational basis) according to an environmental legal framework.

This environmental sieve, in turn, is not at the sole discretion of the sponsor: by exercising its own sovereignty, the Sponsoring State has taken on article 4, paragraph 4, and article 139, paragraphs 1 and 2, UNCLOS III. Thus, the first parameter of due diligence to be respected is added by the written expression "responsibility to ensure", according to ITLOS.

This "responsibility to ensure" is constructed, by ITLOS (2011, p. 41), very analogously to duties imposed on Uruguay in the ICJ - Pulp Mills on the River Uruguay case, when the judgment of this case opposed Argentina and Uruguay on a pulp and paper mill in the Uruguay River basin, in disregard of the Uruguay River Statute, and of article 194, UNCLOS III, As such, ITLOS defines that this preceptive expression contains: (a) an obligation of means, and not of (b) translated into a duty of conduct, c) verifiable by diligence.

This diligence, in turn, must contain (a) a precept in the domestic legislation of the Sponsoring State that; (b) mirror means that ensure and protect, in terms of the state-of-the-art of practice, the other nations of the planet that the CHM of the Area is subject to a preventive regime of the main and relevant environmental risks (ITLOS 2011, pp. 43–44).

This at internal legislation, in turn, is outlined as a form of two other propellers namely: (a) the absolute non-applicability of the CBDR principle when in the light of environmental responsibilities for exploration or exploitation facts in the Area without due care due diligence (or insufficient care) (Lawand and Valdez Silva 2020); and (b) the submission of its internal norm of imposition of due diligence on the sponsored to the state-of-the-art of the practised by the other States Parties worldwide under similar obligations and towards the Declaration of Rio/1992 (ITLOS 2011, pp. 44–49).

With this judgment, under the terms of UNCLOS III for purposes of sustainability as the ordering principle of the sea regime it is not only existing and binding in the colours painted by Schrijver, but it makes the "steering effect" described by Beyerlin and Marauhn (2011, pp. 33–35) of an originally non-binding rule (Dec. of Rio de Janeiro 1992) operating in a new plan (that of the jurisdictional courts), which makes them, at least in effectiveness, binding to the Regime do Mar.

Working with a thought in these terms, and in the drawings established by Schrijver (2008, pp. 288–325). Sustainable development is a principle present and declared in international legal regimes. However, even due to its complex infrastructure component, its shape, parameters, obligations and rights, its binding quality or not, are only available to the interested party when the principle on screen takes on the colours of the regime in which he will work, combining political options with scientific information. In much the same circumstance, despite the CHM manifesting itself in different (systems i.e. Area, historical heritage, Antarctica, celestial bodies), its content obeys the special features of the regime in which it has been requested to be imposed as a legal institution.

For Schriver, there is a principle of sustainable development, generically and setting documented in the Declaration of Rio de Janeiro/1992 but. Then, linked, and effective in complementary normative foundations, according to the regime that requests its application, which gives him the exact measure of what was demanded from the international community of that principle for that specific regime, considering the still fragmented reality of international public law.

Marffy-Mantuano (2011) underlines in turn, that sustainability is a general norm and one of the legal pillars of the Sea Regime, applied in all its activities protection because of the safeguarding of the elements of biodiversity, conditioning the options policies, and the action of the institutional framework inherent to the regime.

Concerning mining on the seabed, the author is categorical by saying that this sustainability at sea is necessarily related to the ecological criteria, which include biotic and abiotic elements. In this way, the depletion of ecosystem resources is not an option politically taken in UNCLOS III: the appropriate option has been for the access to a plentiful CHM in capabilities between present and future generations. In other words, after the degrading operation, access must be guaranteed for those who wish to look for the ecosystem in a specific once geographical location as soon as it is mined.

Scovazzi (2000, pp. 213–220) illustrates the problem that Marffy-Mantuano introduced in the concept starting from the ecologically unique systems of the hydrothermal vents of the ocean floor, the author opposes the Convention on Biological Diversity (Rio de Janeiro/1992 – CBD) and UNCLOS III. The issues posed are: Who is entitled to access this particularly important genetic heritage, and who regulates it? As written by the author,

> Oases of life exist also on the deep seabed, despite extreme conditions of cold, darkness and high pressure. In 1977 it was discovered that many animal communities live in the complete absence of sunlight in the seabed where warm and hot water springs from tectonically active areas (so-called hydrothermal vents). These communities, which do not depend on plant photosynthesis for their survival, rely on the productivity of bacteria able to synthesise organic compounds from carbon dioxide. They use energy derived from the hydrothermal fluid of the vents whose temperatures can reach 1200o C before mixing with seawater. Several species of microorganisms, fish, crustaceans, polychaetes, echinoderms, coelenterates and mollusks have been found in hydrothermal vent areas. Many of them were new to science. Due to their genetic material, these species, in particular microorganisms, present great interest for biotechnology 389. The possibility to use the genetic material that makes hydrothermal vents species able to survive in extreme conditions opens new horizons in the field of genetic engineering and offers prospects of promising economic implications (SCOVAZZI 2000, p. 213).

The author excludes the hypothesis that collection of genetic material is compared with fishing, given the divergent purpose of both, and with scientific studies, for the same reason. As a result, all UNCLOS III fishing rules would be banned from the application (UNCLOS III, articles 116 to 120). The same can be said about scientific research at sea (Part XIII) (Scovazzi 2000, p. 214).

Still, the author explains those statements in article 3, CBD,[8] and article 22, CBD,[9] is embedded in the Law of the Sea regime, regarding the use and discipline of the genetic resources of the Area, in an integration with article 2, UNCLOS III (Scovazzi 2000, p. 216).

And last but not least, as will be seen, the analysis of Case 17 ITLOS—criticism also, the critiques of Wolfrum and Tanaka, below—impose on ISA and the Member States the protection and management of all forms of access to the information. Area—article 157, UNCLOS III, which encompasses ecosystems and their genetic potential. The heritage represented by biodiversity and its forms of exploitation is no exception and must be compatible and consistent with mining activities in the Area.

Considering this, and since it is possible to access this genetic heritage and regulate so much, CHM and sustainability are to be considered, which works on this heritage considering two potential activities—theoretically mining near hydrothermal vents—specifically, that relating to sulphated metal) piles—and access to this genetic heritage.

It is noteworthy that the legal standard of the applicable Law of the sea does not harm but stimulates the mining activity of the Area. Same as the legal norm, a competent court determines that the mining activity in the international seabed is not an exception to the CHM or the principle of environmental sustainability. On the contrary, mineral exploitation is under obligation in conformity and harmony with these two governing principles of the Area.

Still, one issue remains: how to align this mining regime with CHM and sustainability when the activity is not an exception made for both.

Two legal paths are formed being (a) to assume that mining activity is an activity interpreted in the final calculation between development and conservation of systems implying, instead, an exhaustion of natural resources by political option and agreed in global terms by all parties in UNCLOS III; or (b) consider this activity as one more activity in the international seabed, imposing, similar all other activities, a duty of

[8] *"States have, in accordance with the Charter of the United Nations and the principles of international law, the sovereign right to exploit their own resources pursuant to their own environmental policies, and the responsibility to ensure that activities within their jurisdiction or control do not cause damage to the environment of other States or of areas beyond the limits of national jurisdiction.."*

[9] *"1. The provisions of this Convention shall not affect the rights and obligations of any Contracting Party deriving from any existing international agreement, except where the exercise of those rights and obligations would cause a considerable damage or threat to biological diversity. 2. Contracting Parties shall implement this Convention with respect to the marine environment consistently with the rights "and obligations of States under the law of the sea".*

observance of multiple access (albeit a posteriori), which would involve a process of recovery of the degraded area (mining).

Marffy-Mantuano (2011) points out that the alternative "a" is not a sustainable option due to the trans-generational bias of access to resources. Therefore, this makes alternative "b" (RDA) the only possible one.

Talking about the seabed environment in addition to national jurisdictions (as agreed upon in the United Nations Convention on the Law of the Sea – "UNCLOS III", article 1), called "Area", requires that we have in mind, in the category of work assumptions, two variables, namely article 136, UNCLOS III, which categorises the seabed as CHM; and the second, that sea, as a source of natural resources, re-signifying this function in view of the mineral wealth (polymetallic node clusters) of the seabed and the technology able to collect them (Marotta Rangel 2014, pp. 472–479).

In general terms, regarding the Law of the Sea in international law, Tanaka (2018, p. 112) states some key stones of sustainability in application:

a) The need to preserve natural resources for the benefit of future generations.
b) The objective of rational exploitation of natural resources.
c) The use of natural resources based on equity, which means considering the needs of States. and
d) The need to ensure that these environmental considerations are integrated into development policies and plans.

When applying the genus to our species for the Area, we find the political "options" well expressed in their normative texts in the international legal regimes of the Law of the Sea and the common heritage of mankind.

Thus, Wolfrum (2018, p. 108), describes:

a) The economic development regime of the Area is special in relation to the other exploratory regimes of the sea, which denotes the existence of Part XI, UNCLOS III.
b) although separate and directed to the mining regime of the Area (constituting the authority to regulate so much, the International Seabed Authority – "ISA"), imposes environmental criteria related to the procedure and diligence of environmental quality, in evident and exclusively precautionary or predictive.
c) also, the mining activity in the Area is not exclusively regulated by this Part XI, UNCLOS III but, rather, by the entire normative universe that concerns the other parts of UNCLOS III (especially its sources), the Principle of International Environmental Law for Sustainable Development and its versions for the Law of the Sea, as well as the concepts and regulations related to the CHM.

What Wolfrum (2018) and Tanaka (2018) synthesise is that although there is a legal regime for mineral exploration and exploitation in the Area, especially conducted by ISA, it is not the only one, nor can it be interpreted in isolation, especially to determine content ESD specific to the Area.

2.2 Area Administration

In previous topics, we said that the UNCLOS III negotiation rounds opted for the unprecedented template for the management of a common good, the Area. In this chapter, we will see what this option is negotiated and why international society has adopted an unprecedented system.

First-ever, the sequential collective malpractice of the sea, in its free, common, and peaceful use, between the 1960s and 1990s, of the twentieth century. Examples of such malpractices are such phenomena as the flags of convenience, dark ships, the resurgence of pirates and high-impact environmental accidents (with severe damage to the ecosystems and economies) (Octaviano Martins, 2016, pp. 211–213; Goettsche-Wanli 2018, pp. 22–29; Singh 1973, pp. 50–64 and 112–114).

Much of the normative framework that allowed several of these phenomena to recur was repeated in UNCLOS III, given its character of consolidation or to allow greater international adherence, advancing in other points or paving a better path for the future, which was admitted by himself Koh (2016). Tharpes (1989, pp. 611–614), in turn, was categorical in stating that UNCLOS III, considering this same formation of its precepts, meant a loss of opportunity to set parameters for marine pollution criteria, moving forward, even in modelling like MARPOL, for technical emission criteria and command and control systems. However, the same course could not be taken in a novel phenomenon and classified in an express and differentiated way by UNGA itself, that is, the uses and activities in the Area.

Ostrom (1990, pp. 1–18) pointed out that if a common is threatened by the malpractice of its users, equal in law, Leviathan, empowered by its sovereignty, raises its unique club from the monopoly of force and disciplines the issue to maintain the common thing and the benefit/benefit of all, even though this State is protected by a private organisation.

This notion influenced the decision-making of permanent missions to choose a centralised regime of administration for the Area. Issue: there is no Leviathan but several. All of them are sovereign.

In another point of view, from the perspective of the GPGs, the threat to the commons represents a threat to all actors and that the sovereignty regime does not help, and sometimes obstructs, their management.

This root sickness of the formation of the administration of the Area can be described as follows:

- Time 1: sovereignty establishes the area's administration system, consisting of ISA and satellites (such as the company).
- Time 2: ISA (and its complex) is responsible for the government (and governance) of the commons known as Area, whether in accesses, or in the discipline of uses.
- Time 3: in the case of illegal use or that prevents another use of the Area by another party ISA must impose on the party that applied for harm or nuisance the recompositing of the Area.

Systems of conflict-resolution through war aside, it is pointed out that the solution for this regime to the mandatory ITLOS jurisdiction with the members of UNCLOS III, without any possibility of reservation in this aspect.

Thus, we have the limit situation for study purposes: a mining act authorised by ISA creates the impossibility of future use of the mined area by others, which is perfectly plausible considering the state of knowledge of the ocean floor and the volume of mining processes to destroy landscapes, biotic and abiotic structures.

2.2.1 *The International Seabed Authority (ISA) and Its Institutional Mission*

Coming from Resolution I, of the United Nations Conference on the Law of the Sea, although ISA has been present in UNCLOS III since 1982, its current format derives from the 1994 reform and is described in precepts in Section 4, Part XI, of the Convention.

It is an international organisation called International Seabed Authority (ISA) and composed of the signatories of UNCLOS III (article 156 (2)). It was given by institutional mission—purpose—whose purpose is the implementation of Part XI, of the Convention, considering the administration of the Area as CHM, in particular its resources, under the terms of article 157, UNCLOS III.

In accordance with the mission, an organisational structure and management and action powers were also established in Kingston, Jamaica (article 156 (4)) which were also conferred to ISA (UNCLOS III, article 157) for the mission performance. These powers are distributed in organically distributed competences (according to article 158), namely:

(a) The Assembly (articles 158 (a) and 159)
(b) The Council (article 158 (a))
(c) The Secretariat (article 158 (a))
(d) The Company (article 158 (b))

Although it was decided to establish an ISA, an international organisation with delegated powers to manage activities on the international seabed (including the granting, in the name of all nations, of the activity), an international tug of war, strongly influenced by the geopolitical axis of North-South conflict, it placed it as an object in relation to its functions. We explain as follows.

The licence regime did not decrease, but the ISA organisation chart suffered a strong change in its system of power, command, and control of licences

Initial, and as seen, mixed systems of interpretation of the CHM (freedom of access to the Area v. reserved areas of the Area) are still completely subject to the system of grants and licences.

Next, the Assembly missed some of the original 1982 powers on the Council. Among them, we highlight (a) the revision of the Rules for the Use of the Area's

Resources; and (b) sharing with the Council the production of technical rules for exploration and exploitation in the Area.

The Assembly, composed by all the members of UNCLOS III with the right to one vote, must establish the general policies, the election of the members of the Council, the Secretary-General, and other bodies and, in competition with the Council or on recommendation from there, establish rules for the prospection, exploration, and use of the Area's resources (UNCLOS III, articles 159 and 160). It meets at the headquarters or elsewhere, subject to the approval of the Assembly itself.

The Council, a collegiate body of representation composed of 36 members, has executive functions a) conclusion of international agreements—including with the UN);—execution of the policies determined by the Assembly; approval of the work plans and analysis of their pertinent reports from the company or from the contracts sponsored by the Member States for the exploration (and future exploitation) of the Area; recommendation to the Assembly of the technical criteria of these activities in the Area and sharing of financial benefits (UNCLOS III, article 162).

The Secretariat is responsible for the administrative conduct of ISA and performs the functions of the company while it is not yet established (UNCLOS III, articles 166, 167, and 168).

The company is the first international business organisation with the purpose of intervening in the economic mining domain of the Area and depends on the Council's exploration and exploitation orders. Although created by UNCLOS III, it has not yet been implemented. Their field of work will be in the reserved areas of the funds (UNCLOS III, articles 158 and 170).

2.2.2 Mining and Other Economic Activities in the Area

The Area is the object and serves as a direct or indirect theatre of operations for various transnational activities, some of which are fundamental to the running of contemporary society others are consistent in world heritage as much as the international seabed itself, i.e. oil and gas extraction, transportation of oil and gas, bed of communication cables, environmental services—such as scavenging and decomposing ocean waste by scavengers or capture and reserve of carbon from the biosphere—, genetic reserve—even poorly discovered—, reserve and historic heritage site, archaeological and geological features and a site devoted to the scientific research.

ISA, in turn, and in harmony with what we have explained about the historical functionalities of the sea, observes the Area rationally in terms of its economic uses, although all the anticipated uses of the Area are under ISA's total or partial competence, the regulation of this CHM or GPGs was limited to mining activities (Egede 2018, pp. 158–159), which agrees and is contained in the criticism of Scovazzi (2000, pp. 119–121) in the same respect. The Italian author compares

Pardo's speech[10] with Part XI, UNCLOS III, and distinguishes at this a smaller and narrower normative universe of administrative normative application, the reason why ISA could hardly diverge from the cycloptic action alerted by Egede.

Costa de Souza (2019, pp. 53–54) and Araújo Júnior (2017, pp. 375–380) still comment on this emptying caused by the reform of UNCLOS III in 1994, with the inversion of the ISA chain of command and the change in the rules for the transfer of funds mining technology, to be regulated only for opportunities in the market now.

In view of these factors, Egede (2018, pp. 160–170) points out that, again, international regulation has not managed to avoid predatory competition between States Parties, creating "convenience" Sponsor States—"in reference to 'convenience flags").

It is because of the provisions of the Mining Code, in the light of the CHM and sustainability principles, incorporated by Case 17 ITLOS, while there are still gaps to be answered by law and science, the Area is still one of the most regulated geographies in the world over for the purpose of mining.

Nowadays, there are 30 exploration agreements for the Area conducted by 21 entities, none for exploration—mining—, which means that no large-scale mineral extraction activities for commercial purposes can be conducted. It also means that only research activities related to the ecology and ecosystems of these 2 million km^2 can be conducted by sponsoring entities and States (ISA 2020), based on the contents of these contracts.

To proceed from this exploring to the exploration in which the mineral extraction activity for the commercial purpose on a large scale may be conducted, it will be essential (under the terms of Part XI, UNCLOS III):

(a) The approval of the later parts of the Mining Code, with the acts aimed at exploitation being perfectly approved; and
(b) The environmental, geological, and economic studies conducted, consistently.

2.2.2.1 Mining Regimes and Their Code in the Light of ITLOS

The mining regimes in the Area are regulated by Part XI, UNCLOS III, and are conducted under ISA's administrative dictates, embodied in acts currently under construction, called the "Mining Code". That code is anything but that: it is a set of technical acts, voted and approved by the ISA bodies after 1994, with the aim of conducting the technical and control of the work described therein.

Currently, there are four technical acts edited by seven decisions of the Council and Assembly, and one in writing and discussion, which are:

Exploration Regulations:

[10] Speech by the Ambassador of Malta Arvid Pardo on 11/1/1967 that gave rise to Res. 22ª. UNGA 2749 (CHM) and the preparatory acts of UNCLOS III.

(i) For polymetallic nodules: Council Decision 2013 ISBA/19/C/17 and Assembly Decision 2013 ISBA/19/A/9
(ii) For polymetallic sulphides: 2010 Assembly Decision ISBA/16/A/12 Rev. 1
(iii) For ferromanganese crusts rich in cobalt: Assembly Decision 2012 ISBA/18 / A/11
(iv) For questions related to ISA's management regarding exploration contracts:

  (iv.1) At prices and rates for administration and supervision of exploration contracts due to ISA: Assembly Decision 2013 ISBA/19/A/12
  (iv.2) At ISA budget for 2019–2020: Decision of the ISBA Assembly/24/A/11
  (iv.3) The procedures and criteria related to the extension of the work plan approved for exploration (Section 1, paragraph 9, Annex of the Reform of 94): Council Decision ISBA/21/C/19.

(v) In writing, for questions and contracts related to exploitation: ISBA/25/C/37 and ISBA/24/LTC/WP.1

In fact, the mining procedure—theoretically—regulated by CHM, Sustainability, ISA, and the Mining Code obeys the following dynamics:

While each mineral category has its own rule (s), in gender the following sequence of phases is obeyed:

(1) Prospecting notification—exploratory
(2) Submission for the approval of an exploratory work plan
(3) ISA contract signature—proponent (sponsored contract) based on the approved work plan

On the other hand, the exploitation—mining—process for the Area has not yet been standardised. The working ground of the present standardisation will depend on and will be consequent to the exploration results plus the normative demands related to the national standards that are related to the environmental duty of diligence and respect with principles mentioned above. The legal basis for establishing obligations and rights will also be contractual.

However, although the environmental assessments have been minimally structured in its procedure (Annexes II and III of the Technical Study Isa No 16 (ISA 2016, pp. 41–62)), we must, first, it is almost all focused on the character of prevention and/or precaution, lacking process and workflow structures to what, euphemistically, it is called *impact management* (ISA 2016, pp. 41–62).

Next, there was a recognition that procedures must be taken in view of an impact on the international seabed due to mining acts.

Still, regarding this procedure, stakeholders express that the political choice the situation to address exclusively the preventive character of mining activity in the Area is missing deeper scientific elements (ISA 2017, pp. 32–34) that is capable of, at least, providing an understanding of what it is and what is the functioning of the ecosystem dynamics of resources exploited. The maximum that has been achieved so far with the experiments of Peru Basin, was a recovery of the amount of biomass through microorganisms, microfauna, and some polychaetes but not of quality, *i.e.*

the return to the qualitative and quantitative bases of biodiversity that would restore the dynamics and the autonomy of such a system as able to maintain a life in its peculiar forms.

In other words, there is no bright enough baseline to avoid certain behaviours (in time, mode, or spaces). And in that sense, ISA gives in to a prerequisite for the regional work plans in environmental matters (REMP) component of the work plan to be approved by the Council, to a sufficient scientific advance in order to allow metrics of natural systems (ISA 2019, pp. 6–7).

Regarding REMP in the CCFZ, ISA

(a) Does not give a clear idea of what "scientific sufficiency" means with regard to a proposal for mining plans
(b) Recognises that both prevention and polluter pays (and, consequently, the recovery and compensation of degraded areas) can occur in mining plans and operations:

> Knowledge gaps of particular relevance to REMPs were identified in the following categories: • Environmental baselines (including models and validation) for the seabed and overlying water columns in the region, including descriptive-, process-, and ecosystem-services oriented baselines and cumulative impact assessment; • Relationships between REMPs and PRZs / impact reference zones (IRZ) as mandated by the rules and regulations of ISA; • Mining technologies and methods, including seabed and water column disturbance characteristics (spatial and temporal scales of impacts), monitoring approaches, triggers for stop actions; • Stakeholder engagement and roles in development of REMPs; and • Operational definitions of areas at risk of serious harm and means for their protection (ISA 2019).

However, like its precursor study cited above, it explores the concepts of the very first possibility and it ignores the opportunities of the second.

2.2.2.2 Environmental Obligations of Mining Actors in the Area and the Obligation to Recover Environmental Damage

There are three guiding threads of legal thinking that lead us to observe what are the environmental obligations on mining actors in the Area, in particular, the existence of the obligation to recover the portion of the degraded Area in the process, namely, (a) the CHM axis, (b) the GPGs axis, and (c) the sustainability axis.

Although they are only legally logical ways of looking at the same point of interest, and even therefore sometimes overlap in reasoning, occurrences, and deduction processes, it is interesting to look at how the international environmental law opens in institutes (albeit fragmentary) that are complement and end up forming a normative framework of the circumstance that regulates.

We begin with the sustainability axis.

As noticed that Marffy-Mantuano (2011) understands that international seabed mining as another activity supported by the Area, stately, as well as all other activities, a duty of observance of multiple access (albeit a posteriori), which would involve a process of mining degraded area recovery (RDA).

We have also observed, through the same author, that those multiple accesses are permeated by the sustainability pillar and this, in turn, has its management in time mediated by the precautionary principle (Tanaka 2018, p. 112; Wolfrum 2018, p. 108), despite its management of other variables by other principles set out in Dec. do Rio- 92, namely applicable to activities in the Area.

Hence, we have reached a conceptual conclusion that a certain geographic location in the Area that hosts mining exploration or exploitation, given the nature of the mining activity, should be supported by an RDA operation in recognition of the principle of sustainability.

However, it is not only for this reason that RDA is required in this case. Other elements must be considered as fundamentals of the RDA obligation in the Area, by virtue of article 192, UNCLOS III: the concept of multiple uses of the Area, such as CHM; the international due diligence obligation; and the concept of GPGs.

Such reasons are necessarily deduced from the conclusions not only of the ITLOS judgment that now is under analysis but of ISA Technical Study No° 10, which states that licensing regime for mining activities in the Area implies the following responsibilities of the Sponsoring State (ISA 2012, pp. 29–30):

> International obligations The Working Group identified the following obligations under international law as overarching principles that should be incorporated in any statutory framework for offshore mining: Duty to protect and preserve marine environment (Article 192, UNCLOS); Precautionary approach (Principle 15 of Rio Declaration; ITLOS Advisory Opinion; ISA Mining Code); Duty to prevent, reduce and control pollution from seabed activities (Article 208, UNCLOS); Best environmental practice (ISA Mining Code, ITLOS Advisory Opinion); Duty to prevent transboundary harm (Part XII, UNCLOS; ITLOS Advisory Opinion: Rio Declaration); • Duty to conserve biodiversity (Article 3, CBD); Prior EIA of activities likely to cause significant harm (Article 206, UNCLOS); Ongoing monitoring of environmental impacts (Article 204, UNCLOS);
>
> The following principles might also be included: 'Polluter-pays' principle (Rio Declaration); Regional cooperation/integration in monitoring, processing, and capacity-building (Articles 276–277, UNCLOS); Identifying mechanisms of capacity building (Part XI, UNCLOS); • Accountability and transparency (Aarhus Convention).

Concerning the concept of multiple and universal uses of the international seabed, we observe the duty to *No Harm*, expressed in the technical study. This duty, in international environmental matters, became known in the *Trail Smelter* case—that has already been addressed in this work—and dealt with two sovereignties discussing the legality of their intra territorial sovereign acts, even though they generated extraterritorial damage to the sovereignty that lies ahead. The *Trail Smelter* case is supported by the *Papeleras* case, *Gabčikovo-Nagymaros* case, and *Pacific Nuclear Tests* case, all pointing out that it is a nuisance and, as such, an act of international offence with responsibility, aggression against the people or property of one sovereignty, albeit by a legal, peaceful, and sovereign act of another.

So if mining is authorised in the Area, it will be a legal act. If, however, mishandled, it will prevent the other uses of this geographical location by the sovereignty that holds them. And what authority holds them, in the meaning of use, access, or custody? For the concept seen by CHM and the reference of *Bluefin*

Tuna case and *Whaling in the Antarctic* case, all recognised members of the UN. Reason for this is that declaration of universal heritage stems from an act of the UNGA, which was, in turn, repeated and disciplined between signatory parties to UNCLOS III.

In this case, the RDA is an instrument resulting from the restoration of the damage caused by *nuisance*,[11] mandatory given these multiple uses and duty of non-aggression against the other sovereignties.

Although in arrears, Convention on the Regulation of Antarctic Mineral Resource Activities (CRAMRA) has a mining regime that is analogous to that of the Area, a CHM—that may, due to its environmental services, be considered GPGs),—which can only be mined under the regime of recognition of some qualities inherent to the exploiter, being in CRAMRA, the recognition of territorial rights, in UNCLOS III, the licensing regime. In CRAMRA, the obligation of RDA becomes clear when reading Fitzmaurice (2002, pp. 231–232):

> CRAMRA provides that an operator engaging in Antarctic mineral resource activity shall be strictly liable for damage to the environment and implies that the operator is liable for payment if there is no restoration of status quo before for the environment (including residual liability of a State). This form of valuation is subject to criticism on the ground that the costs of restoration or replacement applied mechanically may exceed the value of the original resource.

The international accountability regime has no reason (i.e. there are no provisions in UNCLOS III that contradict it) that this does not apply to the bribe *ubi eadem ratio ibi eadem jus*. In: other words, if there are similar and at all similar regimes, there is no reason not to apply the same provisions. In case of damage by mining in universal heritage, there is no reason for the operator to recover the degraded area to the maximum of the status quo since it is possible.

For their part, references to CBD; article 208, UNCLOS III; and the polluter-pays principle leave doubt as to the use of RDA as a mandatory instrument. It is necessary to recognise, due to the lack of any other alternative or possibility, the duty to restore the Area in RPDA, based on Case 17 ITLOS, article 61 (3), UNCLOS III, CHM, and sustainability.

RDA in this field is a key role in fulfilling obligations. If there is no mining methodology in the seabed other than the depletion of the ground's initial soil horizons, there will be no way out of using RDA methodologies.

[11] "nuisance and the unreasonable, unwarranted and/or unlawful use of property, which causes inconvenience or damage to others, either to individuals and/or to the public. Nuisances can include noxious smells, noise, burning, misdirection of water onto other property, illegal gambling, unauthorized collections of rusting autos, indecent signs and pictures on businesses and a host of bothersome activities. Where illegal they can be abated (changed, repaired, or improved) by criminal or quasi-criminal charges. If a nuisance interferes with another person's quiet or peaceful or pleasant use of his / her property, it may be the basis for a lawsuit for damages and/or an injunction ordering the person or entity causing the nuisance to give up (stop) or limit the activity (such as closing down an activity in the evening)." (Hill and Hill 2021)

Chapter 3
Recovery of Degraded Area

3.1 Degraded Area

The prerequisite for defining recovery of degraded area ("RAD") and its corresponding Plan ("PRAD") is to conceptualise the Degraded Area.

For this task, we will observe the infrastructural technical rule in this respect, since it is more universal in terms of the superstructural legal rule. That is to say: instead of investigating each internal legislation of the sponsoring countries, we will observe the technical rule of environmental recovery, since it is objectively likely to be demanded by other international actors against the degradation agent, according to the understanding set out in the previous chapter.

One of the sponsoring countries for exploration projects, with intention to engage in future exploration of the international seabed, is the Federative Republic of Brazil.

Its proposal and sponsorship are differentiated from the others in the combined mining of the Area for the following reasons in the mining context: (1) the sponsoring mining/operating company is a company owned and controlled exclusively by the State; and (2) the proposal is outside the most harassed areas (which was in the Pacific, especially the Clarion-Clipperton Zone (CCZ)), in the Atlantic (Rio Grande Massif and Basin), in an area subject to conflicting processes (at the International Seabed Authority (ISA) and Commission on the Limits of the Continental Shelf (CLCS)); (3) the Brazilian proposal is not associated with any other State Party to the United Nations Convention on the Law of the Sea (UNCLOS) III.

The peculiarities make it the perfect case for an essay about its environmental rules applied to the process of exploration and exploitation of the international seabed. In other words, how does Brazil understand the recovery of areas degraded by mining on the deepest ocean floor?

Running at an infrastructure level, we have two fundamental concepts of degraded areas:

A. E. Lawand Junior, *Legal Aspects of the Recovery of Areas Degraded by Mining in the International Seabed*, SpringerBriefs in Law,
https://doi.org/10.1007/978-3-031-12492-1_3

Tavares (2008, p. 3), Brasil (2016, pp. 2-1 and 2-2), and Gikuchi (2019, p. 1), are unanimous in observing degraded areas as having two fundamental elements: 1st.) Intervention and/or human activities; that 2nd.) generate harmful or adverse effects in each geographical location in order to condition or alter its possible, planned, or potential uses.

However, ABNT NBR 10703 (a Brazilian technical rule for standards) points out that an area is degraded when there is "adverse change in the characteristics of the soil in relation to its various uses, both those established in planning, as well as the potential".

Exception for the ABNT definition, the above definitions exclude natural occurrences (i.*e*.., the fall of a meteor, a volcanic eruption, a non-standard eruption of hydrothermal vents) that may cause the effect of altering the soil for a given anthropic purpose, such as, it also links it to effects that curb uses, within planning (*i.e.* technological) perspective, according to the state-of-the-art at the time of the analysis of degradation.

We realise that this technical concept is not applicable insofar as options for using the Area draw up access and multiple uses, especially mining, pipelines and cables extended on the international ocean floor. The ABNT definition does not incorporate the concept of human intervention, which is valued in UNCLOS III and the ISA Mining Code.

Let us then apply the concept chosen to the Area.

The Area, due to the special activity control regime resulting from its principles and rules (see Chap. 1), has activities regulated by ISA.

ISA, for its part, issued guidance books, as well as, under the terms of the relevant Mining Code, EIA, and other informative activity studies (2018), possible environmental impacts that are created by mining in the area, before authorising its exploitation and regulate it—in the part of the Code not yet approved and under discussion.

It is important focusing on this point for the following analysis.

The largest biome in the world, represented by the polygonal of the Area, carries the title of being the least geography by those known to human beings. In three recent editions and with significant impact of collections about knowledge representative of the biome of the Area (Wedding et al. 2015, p. 144; Taylor and Roterman 2017, pp. 4872–4873; Santos et al. 2018, pp. 1506–1507), these are all categorical in informing that the volume of knowledge obtained is still fairly superficial.

It infers at this moment, in the very notion of degraded Area, how to figure out the present and future potential uses of the Area if I do not even know it properly?

It is an issue that touches on the concepts behind and the discussions about the (legal) principle of precaution. Understand why:

All human activity can, at an economic level, be reduced to the micro-economic formula of production. In this formula, regions and their biomes are raw material (Derani 1997, pp. 72–73), while human transformation activities are concentrated and ordered in the capitalist production unit (UP).

Throughout this UP, the predominant factor of possibilities, accesses, and production capabilities are determined by technology (Piketty 2013, pp. 71–99).

It is not due to a deficit in the current state-of-the-art that biomes are explored, as there is plenty of technology and their development reaches levels that have not been historically matched, reaching the point of becoming a fundamental element in motivating governmental decisions, as a way of legitimising them or legalising it (Giddens 2000, pp. 42–45).

Therefore, we have the UP variable. What we do not have is a variable of knowledge of the raw materials which exist there, of the potential of the biome due to an exploratory lack.

ISA admits this when recommending that the exploratory concession area in the CCFZ bordering particular regions should be approached more carefully—in terms of non-exploitation or minimally impacting exploration—in view of the fact of the considerable knowledge gap on the environment located there.

In the absence of this data, one of the variables of the microeconomic- equation falls, the totality of uses is unknown, not to mention the potential for the future, for the purpose of multiple accesses—according to the Regulatory regimes of the Area they value). Therefore and for this reason, there is no way to measure, technically and properly, a degradation that meets the necessary technical-scientific concept—*i.e.* degraded area—to fulfil the international obligation (RDA).

With the absence of the concept of determining the complete degradation of the area, let us observe the precept and the norm that derives from it.

Precaution/prevention, at the international level, is defined as follows:

> Declaration of Rio de Janeiro-92. Principle 15." To." protect the environment, the precautionary principle must be widely observed by States, according to their capabilities. When there is a threat of serious or irreversible damage, the absence of absolute scientific certainty should be used as a reason to postpone effective and economically viable measures to prevent environmental degradation.

This definition was, in turn, interpreted as follows:

> The precautionary approach aims to take preventive measures to respond to probable or potential risks which cannot be objectively identified through present-day science, but which might create environmental damage in the future. In this sense, the precautionary approach is intertemporal by nature. It can be considered, therefore, that a time element in the precautionary sense is incorporated into the sustainable management of marine natural resources through the application of the precautionary approach (Tanaka 2018, pp. 116–117).

For this definition, it is understood that the scientific uncertainty of the risks related to a given activity on the elements of the biome to be explored—another name for "natural resources"—does not have sufficient reason to prevent such proposed activity, whether, due to its licensing, other care and risk controls and externalities are taken or future damage cannot be reasonably seen.

Note that the Precaution has some axes of occurrence and work (a) one has a notion—at least sufficiently solid—of the biome to be accessed—explored—and cannot say what the UP—due to lack of technological information—will do with this biome, or (b) the perception of total and traumatic annihilation of the patrimony already conquered/preserved is much more serious and has much more value than any gain that the operation of the UC can provide (Schrijver 2008, pp. 349–358).

In this case, both can be considered, one does not have the notion of the components of the ecosystems of this biome. The UP is well-developed from wire to wire (Miller et al. 2018, pp. 9–11)—which was even described in Case 17-ITLOS (ITLOS 2011, pp. 45, 47 and 54)—but it is not known what this process does will reach (Wedding et al. 2015, p. 144), which heritage will cause damage. The "externalities" that it will generate are unknown. It is unknown, either, if there is any risk of total and traumatic annihilation of the heritage already conquered/ preserved, not by technique but by not knowing where it will be applied.

Actually, ISA attempted an RDA approach based on a desktop-control area plan, or RDA by comparing the surrounding fragments.

Such a scheme was made up of establishing areas of anthropic operation and conservation areas with divergent functions. Among them, buffer zone between the area of anthropic operations and study reference for baseline measures for prevention and environmental recovery (Wedding et al. 2015, pp. 144–145).

However, the gathering of recent data on this RPDA pointed out that the central hypotheses about the performance of seabed ecosystems have not been confirmed, and the restoration of the ecosystem has not been achieved within the intended period. Which necessarily means a shortage of sufficient knowledge to draw a baseline or, still, a recovery objective (Santos et al. 2018, p. 1506).

Technical standpoint, this is the case of applying the Precautionary principle in its original textual sense from Dec. do Rio-92.

It was also important to say that the instrument of Precaution par excellence, the Environmental Impact Study (EIA) is the instrument that uses all the volumes of the ISA Mining Code to assess environmental impacts.

We have seen, by logical analysis, that Precaution will not be legally sufficient. An EIA that relates an international seabed mining UC proposal to a seabed with several scientific gaps in information will certainly not be as efficient.

To demonstrate the hypothesis, it is necessary to observe the legal and technical reference of this EIA.

3.1.1 Mining Area EIA

The Mining Code and the coming exploitation rules must be restricted to the art. 147 and 235, UNCLOS III. As such, it is necessary to note that in addition to the precautionary and preventive precepts, the remedial precepts in a mining context are not only necessary but will give meaning to the CHM and GPGS system, as well as to the Sustainability required by Case 17-ITLOS, in interpretation to the art. 304, UNCLOS III, cc. Declaration of Rio/1992.

There is no exploitation or mining technically or procedurally codified or prescribed within the scope of ISA. Technical Studies 21, 22, and 24 give an account of the degree of scientific uncertainty as to the topic.

However, any exploration activity in the Area must be preceded by an EIA. In turn, ISA (2016) signalled that the EIA's term of reference will be based on the practices already enunciated in Technical Study N°. 10, referred to above.

The privilege given to Precaution and Prevention in relation to the Polluter-Pays does not mean, in any case, an abolition of obligation to recover. As already seen, RDA is an obligation of the mining regime resulting from the mandatory interpretation of the distinct parts of UNCLOS III. In turn, RDA and its methodology are also instrumental in preventing degradation in areas other than the area directly affected by the activity in question.

3.2 The Recovery of Degraded Area

Tavares (2008, pp. 5–7) and Hiddink et al. (2017, pp. 8303–8304) describe RDA as the planned set of anthropic actions aimed at recovering the degraded ecosystem in each area and making it stable (in its dynamic balance) from its own characteristics for a given purpose chosen or the return to the status quo ante of environmental quality.

Across more detailed words: a plan will use environmental engineering techniques and related—or overlapping—sciences for restoring ecosystem dynamics.

The methodological requirements for RDA operations are dependent on which biome applies. However, and even because of the didacticism in which it was put, we will bring here the requirements in articulation of forest RDA that, mutatis mutandis, will be methodologically extremely useful for us:

A. Conserving biodiversity means recognizing, inventorying, and acting, aiming to keep these differences to the maximum, that is, to maintain the greatest variability of living organisms, communities, and ecosystems, in order to meet the needs of present and future generations. [. . .]
B. It is not possible to start a forest restoration or recovery process in a degraded area, without first considering: (a) whether the species to be planted are of regional occurrence; (b) the micro basin as a unit of analysis; (c) the causes of the degradation; and (d) natural succession processes. [. . .]
C. The current "models" and the tendencies to recover a degraded area, conserving biodiversity, must be implemented based on a set of experimental techniques, replicable and based on ecological processes, which also consider the socio-economic aspects. [. . .]
D. The state must be the driver of public policies for the reforestation sector, aiming at the conservation of biodiversity, in addition to providing information and "tools" appropriate to the forest restoration process. [. . .] (Barbosa et al. 2008, pp. 7–11).

Keeping in mind that the requirements A, B and C, above, necessarily imply in-depth knowledge of degraded ecosystems for any RDA plan. We go further, the stated requirements demand knowledge from the RDA planner of the trophic and genomic relationships between biotic elements (*i.e.* evaluation of successional stages of the biome on-screen and relationships between biotics and abiotics as a way of restoring and/or increasing them in the geographical to the point that, harmoniously, sustain and reconstitute the original ecosystem.

These requirements are methodologically important because (a) they will enable the construction of what was the area and/or the ecologically balanced Area—or "baseline",'' or "LB");''—(b) state, from the LB, the magnitude—albeit by approximation criteria—of the damage—in quantity and quality—suffered by the Area; (c) the planned and harmoniously coordinated intervention measures to reverse the damage to the BL again.

At that point, we are going back to the infrastructure in bibliographies.

ISA is in full discussion of approving the regulatory part on the exploitation of the Mining Code. In a note from LTC/Council/ISA ISBA/25/C/18, published on 19 July 2019, the following can be read:

> 21. Draft regulation 26 (Guarantee of Environmental Performance). Considering the Council's request to develop the requirements for these guarantees, the Commission considers that a more in-depth debate with relevant stakeholders is necessary in order to advance the content of this draft regulation, in particular the objectives and requirements of a closure plan. After that, the regulatory text can be updated, and the guidelines developed.
>
> [...]
>
> 27. Draft regulation 45 (Development of environmental standards). This new draft regulation defines the thematic areas for the development of environmental standards. The Commission considers this a reserved space pending further discussion at the workshop to be held in Pretoria in May 2019.
>
> 28. Draft regulation 46 (Environmental management system). The Commission has included the requirement that an environmental management system be implemented. The details of such a system, together with relevant references and principles, should be defined in guidelines.
>
> 29. Draft regulation 47 (Environmental Impact Statement): In response to a series of requests from interested parties, the Commission has reintroduced the requirement for an environmental scoping phase, although the need for a specific environmental risk assessment has been impact assessment process maintained as part of the environment to focus the environmental impact statement on important impacts. In response to comments from stakeholders that indicated some confusion regarding the various elements of the environmental impact assessment, the text was also revised to further clarify the process. The detailed requirements for the scoping stage, including associated processes, should be detailed according to the exploration regime. ISBA/25/C/18 19-05030 7/10 30. Draft regulation 52 (Performance evaluations of the Management and Environmental Monitoring Plan). Although the text of this draft regulation has been retained, changes have been made to reflect that the frequency of performance reviews will now be determined by reference to the period specified in the environmental management and monitoring plan. In addition, the draft regulation now includes an obligation to report to the Council, including any recommendations from the Commission. changes have been made to reflect that the frequency of performance reviews will now be determined by reference to the period specified in the management and environmental monitoring plan. In addition, the draft regulation now includes an obligation to report to the Council, including any recommendations from the Commission. changes have been made to reflect that the frequency of performance reviews will now be determined by reference to the period specified in the management and environmental monitoring plan. In addition, the draft regulation now includes an obligation to report to the Council, including any recommendations from the Commission.

31. Section 5 (formerly the Environmental Responsibility Trust Fund, now the Environmental Compensation Fund). The Commission considered the general feeling among stakeholders that the fund's objective should be restricted to the provisions of the draft regulation 55 (a). The Commission believes that further discussion on the range of financial instruments that should be put in place to create incentives for environmental performance and provide adequate compensation mechanisms under Article 235 (3) of the Convention is warranted. The Commission considered that, although the principal (*i.e.*, capital) of a compensation fund could be delimited and limited to any environmental liability gap that may arise in the future.

Despite the fact that ISA has the official discourse that mining on the international seabed is possible, the stakeholders, in a workshop in Pretoria in 2019, which agreed to emphasise features that are not yet fully defined for the purpose of a fully secure exploitation mining code, even legally.

In the analysis of the items addressed, it appears that the seemingly legal impasse stems from inconsistencies of a technical nature, technical environmental paradigms in response to the required practices usually conducted by the States Parties; parameters of environmental compensation/repair served by the fund; technical environmental rules for a closing plan—closing or decommissioning—of the underwater mine, all of them depends on an eco-systematically designed LB as an assumption.

It is particularly important to note that in the current state of the art, it is impossible to trace BL and, consequently, measure damage or plan measures. The LTC statement is yet another manifestation.

Vonnahme et al. (2020, pp. 4–6) point out data on RDA tests in the Area in the program "Disturbance and Recolonization" (DISCOL).

The DISCOL impact study began in 1989 in the Peruvian basin of the South Pacific at 7° S and 88.5° W. Evaluations took place in 1989, 1992, 1996, and 2015. During DISCOL, the impact of mining in the Area in the benthic ecosystem was the simulated plowing 11 km^2 on several occasions over the seabed at 4150 m. depth multiple times.

Again, a few or nothing—including polychaeta fauna and betonic phyto-physiognomies—has been recovered and the dynamics of the seabed ecosystem have been severely altered, by combining pedological analyses with the traces of betonic ecosystem indicators—porosity, oxygen, carbon and nitrogen cycles, cell count per handful of sediment, etc.), significant–substantial systemic changes were noticed, when compared to the reference area (intact) of the funds, especially concerning bacterial behaviour and its multiplication, consumption and deposition of new substances.

The relentless result of applied ecology makes it necessary to recognise that RDA experiments, in particular DISCOL, are still at their dawn for the Area.

The immediate question here is, is RDA an international legal obligation for seabed mining, however, if its fundamental methodological elements are not currently feasible, how is this rule fulfilled?

Chapter 4
On the Apparent Material Antinomy Between Compliance with Mining Obligations in the Area and Mining Exploitation Rights and Commitments

In Chap. 2, we observed that in the focus of sustainability, the peaceful and multiple use of a legal complex called Ocean, mining in the Area could legally take one of two legal paths, as follows (a) assume that mining activity is an understood activity in the final reckoning between development and preservation of the systems, implying, yes, in an exhaustion of natural resources by political choice and agreed globally by all parties in the United Nations Convention on the Law of the Sea (UNCLOS) III; or (b) consider this activity as one more activity in the international seabed, and, consequently, in the oceans, imposing, like all other activities, a duty of observance of multiple access (although *a posteriori*), to present and future generations, which would imply a process of mining degraded area recovery (RDA).

Through the rights and access guaranteed to future generations, we saw, in the same chapter, that Marffy-Mantuano (2011) points out that the alternative "is not feasible".

Through Case 17 of the International Tribunal for the Law of the Sea (ITLOS) and reading Part XII of UNCLOS III, Tanaka (2018, pp. 112–113 and 115–117) summarises the comprehension as binding environmental sustainability in any activity in the Area.

It makes alternative "b (RDA)" the only feasible and binding alternative.

It appears that in Chap. 2, due to the analysis of the RDA methods that are currently used, as well as the critical analysis of the collection of infrastructure references and sources of ecosystems and their recovery in the Area, we notice that today there is no way to fulfil this legal obligation. Does it mean that the obligation is impossible and is no longer practised?

In other words, what does Part XII advocate and the United Nations General Assembly (UNGA) guarantees common heritage of mankind (CHM), does Part XI stimulate in the opposite direction and in a conflicting sense? Is a legal regime of autophagic and antagonistic international law possible precisely as a result of the bias of international environmental law?

A. E. Lawand Junior, *Legal Aspects of the Recovery of Areas Degraded by Mining in the International Seabed*, SpringerBriefs in Law,
https://doi.org/10.1007/978-3-031-12492-1_4

The analysis of the cases *Whaling in the Antarctic* (Fitzmaurice 2002), *Southern Bluefin Tuna* (Gautier 2016), and *ICJ Belgium/Senegal* (2012) (Tanaka 2018, p. 128) reveals that the perspective of materialising a given activity is not in rights to the regime—mining in the Area—owned by a given actor and its consequences—*i.e.* RDA). Nonetheless, yes, in the combined regime between the precautionary principle and the victim, in fact or potential, of an international environmental occurrence, since in the three cases the material analysis of convenience and opportunity for preliminary injunctions of the invoked right was legitimately conducted.

So we are talking about environmental damage and environmental liability in international matters.

If we observe the production process of international seabed mining (ISA 2012, pp. 12–16; Freidah 2019), the lack of knowledge about this biome and its ecosystems is:

(a) So brutal that even a baseline to identify both status quo ante and post-dredging damage is not viable, scientifically, and technologically
(b) Attempts to isolate and recover test areas (Peru Basin) (Vonnahme et al. 2020, pp. 1–14) have not been positive, with the scarified area also having:

  b.1) Maintained the damage
  b.2) Altered the biomass gene flow at the site

According to these data above, could we confirm that there is antinomy between agreements?

If multiple uses are linked to the biodiversity of the international seabed, we will have UNCLOS III, Part XI, in apparent conflict with the 1992 Convention on Biological Diversity (CBD).

And we could extend the list of issues to be worked on in the international seabed, in harmony with Scovazzi (2000, pp. 124–128, 147–155, 206–212, and 213–219) and considering other uses and means of access.

In conflicts such as this one, between binding norms of the same international category (even if they are integrated and/or complemented by custom), we would apply Vienna-I-1969, notably articles 30 and 41, as a way of solving the apparent conflict of international standards. In this case, notably due to the determination of the intention of the contracting parties and the hermeneutical principle of *lex specialis derrogat lex generalis*.

The apparent problem is the antinomy ends in the same normative body, namely, UNCLOS III.

If multiple uses are linked to the environmental services provided by the international seabed "known and to be discovered" and the intrinsic value of the environment given by each culture and recognised in the international environmental standard, we will have a conflict between UNCLOS III, Part XI, and UNCLOS III, Part XII.

It is for the reason that UNCLOS III, condenses despite the fact that shrinks normative precepts dated from 1982 and 1994, undergoing which agreements are to be subjected to mutations at the normative level (not preceptual) due to the phenomenon of interpretative evolution of the treaties (Tanaka 2018, pp. 117–119) which, as if evidenced by Dworkin analysis, are going through mutations *passi passu* to social changes (gender infrastructure).

In our case, Case 17 ITLOS incorporates, as elements of interpretation to the precepts of Parts XI and XII, of UNCLOS III, the dictates of the Declaration of Rio/1992.

It is in the scrutiny of the legal relationship mediated by the damage and its concept that will conduct us to understand the responsibilities and the role of the RDA in the mining of the seabed and the legal way of solving this apparent intestinal antinomy to UNCLOS III: the apparent antinomy is not at the pragmatic level of UNCLOS III, but at the normative level, hermeneutic later interpretive processes.

Therefore, it is the reading of the values in the time axis that will provide an apparent antinomic read between the norms of the system or will cause the sea regime to balance itself internally and harmonise with the other international legal regimes. The following can be said:

(1) Environmental responsibility, given its trans-disciplinary characteristic, occurs *ex delicto* and sine delicto, which is an exception.
(2) Responsibility *ex delicto* was defined by the International Law Commission (ILC) as a "wrongful act by a State [which] entails the responsibility of that State".[1] The responsibility arises from a breach of a primary, a major source of international law: a binding treaty or a custom in the field of international environmental law, considering the deemed International Court of Justice (ICJ)—Gabčikovo-Nagymaros. Thus, and in this analysis, Fitzmaurice (2002, p. 205) and Douhan (2019, pp. 1–2) agree to point out that:

(a) The liability and, consequently, the damage, are necessarily measured according to a mandatory breach of primary international source.
(b) In view of the fact that there are no concrete predictions or grounds for sine delicto liability, international liability for environmental damage has not evolved (at least frequently) for this modality, except for extremely dangerous activities and as such declared.

The sine delicto responsibility, for its part and soon to be digressed, is what does not require a customary or contractual breach: through the simple assumption of the risk of an activity, the State is caught up in the consequences arising from it.

By applying a generic international environmental concept to the species, *i.e.* the UNCLOS III Area and regime, deals with the definition of terms in Case 17 ITLOS (2011, pp. 22–24), when it demands a mandatory infringement put to infer the State a responsibility and a duty arising from that responsibility.

A mandatory violation of this nature is regulated by article 60 (3) of the Vienna-I-1969 Convention and, as such, alternatively requires:

[1] ILC draft. Article 1 "[e] very internationally wrongful act of a State entails the international responsibility of that State".

(a) In a rejection of the treaty not sanctioned by this Convention; or
(b) Violation of an essential provision for the achievement of the object or purpose of the "treaty"[2]

The issue is that mining in the Area is a legal and legitimate act, according to Part XI and the 1994 Reform of UNCLOS III. However, it is impossible to achieve without a damage load on the marine ecosystem that requires a degraded area recovery (Freidah 2019), under the risk of lasting damage to the marine environment that violates the multiple uses of the Area (CHM—139, 140, and 141, UNCLOS III), the rights of future generations to the potential of the Area (principle of sustainability, Rio Declaration-92 cc. articles 145, 146 cc. 139, UNCLOS III), and obligations of Part XII, UNCLOS III. As already shown, there was no baseline for proceeding with any RDA in the Area.

The antinomy is not between treaties but within UNCLOS III. And it can only be resolved through an international legal norm.

For this reason, we began to analyse temporal perspectives of stabilisation of the legal system of Law of the Sea according to the values that feed its consequent legal norms.

For that, allow us to observe the mining exploration and exploitation process of the Area: in microeconomic terms, can describe the operation as centred on the entity.

It is the entity that will play the role of capitalist production, with the ***input's*** microeconomics—of capital and raw materials that is made up, at least partially, from the requirements of the capital market in terms of mineral commodities and the ecosystems of the Area.

Therefore, these two variables will make up the nucleus of analysis of the weaknesses of the area's legal mining regime in the light of environmental obligations, the capitalist production unit of the mining activity "UP" and/or entity. The source of raw materials, the marine environment of the Area.

RDA is a must when mining the Area. And it is as a consequence of the observance of sustainability, GPGs, and CHM, whether isolated or jointly interpreted on the same fact of environmental interest (mining).

Let us elaborate this, starting with CHM.

[2]Enacted in Brazil through Decree No. 7.030/2009.

4.1 In the Light of CHM

4.1.1 Right of Access (Mare Liberum) or Right of Distribution (Res Commune Omnium)?

The current International Seabed Authority (ISA) Secretary-General, Michael Lodge, speaking in the "Lecture Series" of the United Nations' (UN's) Audiovisual Library of International Law, makes two assertions about Part XI, UNCLOS III, which is of interest in this work (Lodge 2020).

First one, that Part XI is the largest, most complex, most contradictory and obscure of the UNCLOS III Parts, and despite the fact that it deals with all the resources of the Area, it is notably focused on the mining of funds.

Another is the 1994 reform, which gave a new reading to Part XI and the ways of accessing the Area and its resources, considering a continuity of the North-South geopolitical conflict.

The importance of such notes in that lecture is the reason that underlies them, what is the importance of regulating the international seabed in a different regime in relation to the water column that sequences it and why to structure an entire legal regime on a concept that, although it has a definition, does it admit several genres and content that form *ratione materiae* (CHM)?

The signs can be found at two phases of the formation of UNCLOS III: Pardo's speech at UNGA, the embryo of the Convention, and the 1994 reform that brought it into force.

According to Scovazzi (2000, p. 120), the concept that Pardo planned for the international seabed CHM did not come into effect: well, although the "first come, first served" principle predominates in international waters—with inherent limitations—has been put aside and the option for the word heritage "in the sense of leaving property to a future heritor" has triumphed over the word property (UNCLOS III, article 136), and even then some aspects of UNCLOS III reflect a partial distribution of products between developed and undeveloped.

If we take the form in which the capitalist production process develops (Singer 1975, pp. 12–88), equalisations relocation balances in UNCLOS III, in 1982, were in:

(a) Raw materials: the establishment of areas reserved by ISA for exploration on its own initiative or by the company, without appropriation of the ore mined by a specific nation or its consortium
(b) Production unit: technology transfer
(c) Results: the division of results from mining, in a fixed percentage

This was evident in the treatment of technology transfer in Part XI (article 144), when in 1982, the mining tract of the Area was looking for a new model of more equitable development, either by dividing the fruits of the production process, or by equalising production capacity through a compulsory technology transfer between stakeholders. This rule was the pivot for a stampede of countries in the North (industrialised/developed), an act that tacit a serious risk of the treaty not even being in force, much less having the strength and international legitimacy proposed ab initio in contrast to UNCLOS I and II (which did not even enter into force) (Treves 2012).

Circumventing the issue, the 1994 reform, in addition to changing ISA's internal structures and competencies, changed the terms under which international seabed mining would take place. Among these changes is making technology transfer no longer mandatory.

Though, the few that was said about the "negative externalities", which, as previously seen, are misallocations of raw material in the economic process, few were discussed, consigning, at first, to an only apparent discussion of article 139, UNCLOS III.

As in Chap. 1, and by the unified interpretation of UNCLOS III, we can understand that according to the article 145 of the Convention, must necessarily be harmonised with Part XII, ditto, and the polluter-pays principle. What provides us with this direction is the very orientation of article 145, which turns into occurrences already consummated, *ipsis litteris*: "the harmful effects that may result from such activities".

An imposition of non-appropriation, with a distributive licensing system, with an obligation to recover degraded areas and harmonise with other activities (article 147) is significantly different from other ways of treating other activities conducted in the ocean environment.

Thus, access to the international funds has regulated discipline, which differs from mare liberum. But what about the res commune omnium international seabed?

As described in Chap. 1, Cançado Trindade qualifies the Area of *res commune humanitatis*, due to (a) its forms of profit; (b) its non-appropriation by any sovereignty; (c) the condition of legacy to future generations.

Allowing us to take the idea to the limit, a question may be asked: If sovereignties disappear and become other sovereignties, does the nature of the Area remain?

The answer is yes because of these two points of view: the succession between States and what we will call "Delivery".

From the point of view of succession, which differs from "continuation",[3] the rules of international law relating to the succession of rights and obligations between States determine that the assumption of a territory and people, even with other objectives that give rise to a sovereign nation, imply the assumption of commitments to other nations, for political reasons (*i.e.* recognition of this sovereignty as such by other nations, in a "bargaining" sense). For the legal reason of the succession criterion *lato sensu* of rights and obligations (which can only be ignored through an act of denunciation of the treaty) (Stern 2000, pp. 117–175).

From the point of view of "Delivery", the sovereignties of the planet, in their own definitive act, at UNGA, delivered the raw material, *i.e.* space still on the earth's

[3] On the difference between "succession" between states and "continuation" between them, see Stern (2000, pp. 39–47).

crust, one of the fundamental elements of the sovereign State, *i.e.* the territory, with no sovereignty, but with another actor in the international community: humanity, the collective of individuals in the process of time.

It means that only this actor, in this organised manner yet again, can give up this right that was conferred on by the full rights subject of public international law.

In this way, no emerging new State, "like the Democratic Republic of Timor-Leste", or already existing one, "like Portugal", can legally retake what it has been lawfully deprived of: the possibility of seizing the seabed.

Therefore, a second origin of the CHM is evident: a legal regime that permanently prevents the sovereign appropriation of territories and resources by other sovereignties. Therefore, the regime based on the mare liberum principle, which presupposes the appropriation of free resources and free (peaceful) access, is discarded in the Area according to this legal root.

4.2 In the Light of the General Regime of Public International Law

There are two relevant points in time to consider the recovery of degraded Area: the process of collecting ore from the first horizons of the seabed—i.e. dredging—and operations that involve bringing the ore to the surface through the water column.

Although there are two different regimes consolidated in the same diploma (*i.e.* UNCLOS III), Treves (2012) accurately reveals the transversality of the theme—even by reading the judgments of his presidency, especially Case 17 ITLOS—observing that the acts and interventions in the Area impact on the water column.

It is important to understand how legal nuances matter to combine two legal claims that, at the present time, are challenging, but make it a legal entity one in the case thought of in time.

Next, we will explore legal institutes that regulate international public law applicable to the Area.

4.2.1 Moratoriums

Article 305 (1, "c" and "d"), UNCLOS III, allows (and to consider the ISA-OSPAR-2011 protocol even encourages) its parties to associate in parallel, taking the Convention as a normative framework, for compliance of her dictates.

As such, five of the Arctic nations (USA, Russia, Canada, Denmark, and Norway) came together in May 2008 to establish a regime of sustainable and non-exclusive exploitation of this region (since global warming and the melting of the northern polar ice cap opened up new navigation routes and possibilities for the occupation of the region). New routes and science allocate oil, gas, and fish reserves to a considerable extent. In this way, they made the Declaration of Ilulissat-2008 positive in international law (Hossain and Morris 2017, pp. 3302–3316).

By the provisions of this Declaration, it deals with:

(1) Recognise UNCLOS III as a normative framework—including the United States, recognising its customary portion of origin and the CHM—, although they consider it insufficient and in demand of complementation to deal with the pressures and developments in production, trade, and protection of the marine environment
(2) Necessity for activities in the Arctic to develop with adequate ecosystem management, so that there is no depletion or indelible damage, which prevents other Arctic countries from also having this portion of the sea. According to the Ottawa-1996 Declaration—this one, signed by the eight Arctic States, the five mentioned above, plus Sweden, Finland, and Iceland)

Based on that Declaration, and on the subsequent and unilateral manifestations of the parties, Hossain and Morris (2017, pp. 3519–3671) discuss alternative instruments to manage the environment, especially biodiversity in the Arctic having inevitably come up against the alternative of marine areas "MPA" and the option of treaty systems similar to the Antarctic Treaty/1959.

This treaty, in turn, was suffering from the international legal phenomenon of the moratorium. For a moratorium to occur, following necessary tests:

(a) A conflict between subjects of international law; what
(b) Be insoluble under the same space-time but potentially soluble in others

Ilulisat/2008 had its spokesperson for bad news in Russia: geopolitical conditions—*i.e.* superstructural, but in support of the law—do not allow, at this moment in the world and in history, a combination of interests other than the moratorium on everyone's interest (Hossain and Morris 2017, p. 3397).

In our case, the contradictory circumstances are of a legal nature: both the legal values come from the same regime. The mining, as stipulated in Part XII and Annex III, of UNCLOS III. Potential uses universally distributed and the intrinsic environmental value established in Part XI and UNGA acts.

Mutatis mutandis, the observed by the Declaration of Ilulissat/2008 extends to the moment discussed in this topic: the relevant activity of one is not possible to appropriate the oceanic portions of all, especially the Area.

In the case of Arctic nations, the impasse lies in geopolitical considerations that prevent the full advancement of activities at sea in the region when it is taken unilaterally. In the case of the Area, it is noticeable that the mining activity has no way of—yet—rebuilding a global and trans-generational- heritage.

Note that the legal institute of the moratorium does not eliminate rights or obligations from the legal system however rather conditions them in time, suspending them in whole or in part, but recognising them, in consequence, existence and potential application in the future conditions which allows them to observe the terms of effectiveness.

So, it is with the European Convention on Human Rights (ECHR). Note that the moratorium declared on Protocol 6, of the ECHR, imposes on members of this regional system of human rights the moratorium on the exercise of the State in executing certain sentenced to death according to the characteristics peculiar toward those sentenced since 1994 (Aust 2010, p. 221).

The power-duty to plow and conduct the death penalty has not been abolished but has been conditioned in time—and in the European space—and according to the characteristics of the convicted person (*i.e.*. pregnant women or children under 18).

The same phenomenon can be observed in relation to the moratorium on commercial whaling imposed based on the Washington-1968 treaty: the commercial form of whaling, which is even provided for in the diploma is not overlooked. However, its conditions are on time and space (Foster 2012, p. 88).

Applying the concept to our case, the moratorium would not abolish the mining rights of the Area, but it would condition them as time goes on to a scientifically measurable circumstance of building a baseline added to the minimal precision of anthropic interventions as a way to recover the degraded Area for mining. This measure would combine the options of the sovereign nations that signed Montego Bay-82 and declared the Area in UNGA as CHM not as antinomics, but as harmonics in a three-dimensional perspective.

Nevertheless, another problem arises: moratoriums require a declaration.

Binding declarations of a moratorium on a right or obligation that are also binding must share normative categories that can relate to each other at the deepest level of interpretation: which of the ability to overturn one another (articles 30 and 31, Vienna-I-69).

At this point, the same sovereign powers that led to the precepts must be applied to the act of moratorium.

When observing the Antarctic Treaty/1959, the same source of sovereignty—i.e. the States—which set up the exploitation moratorium of the southern continent, established the exploitation possibilities, albeit in a generic and indirect way (article 2 (2 and 3), of the Treaty). In the case of mining in the Area, this possibility was geopolitically depleted in 1994, when the foundations of sustainability and transgenerationality were in place and launched in 1992—or since 1988, if we count the date of the Our Common Future Report. Still, is the moratorium a systemic way out that can make sense of the sea regime?

It can be answered as yes, from the point of view of the articles 160 and 162, UNCLOS III: the forces of the ISA Assembly and Council.

Equally one body and the other exercise the powers of the subjects originating from law who developed UNCLOS III by representation (Aust 2010, p. 183): the subjects of international law determined this international entity known as ISA as the holder of some authority sufficient to declare a moratorium on issues relating to mining in the Area (either directly or formally, indirectly, but with the same effectiveness).

This situation is not something new in public international law. In fact, it is even emblematic and has never been questioned when viewed in the light of the case *Whaling in the Antarctic - ICJ*, previously described.

The case involved Australia and New Zealand versus Japan. In this case, the three countries acknowledged the moratorium on commercial whaling. For its defense, Japan claimed that the Japanese Whale Research Program under Special Permit in the Antarctic (JARPA) was a hunting programme for scientific purposes—which, in theory, would be allowed and not covered by the moratorium. What concerns us is who commanded this binding moratorium: it was not the assembly of all signatories to the Washington 1948 Treaty but the International Whaling Commission, the executive body of the Treaty, with delegated powers, which does not bring together all the signatories.

Therefore, the moratorium, although it is not foreseen expected, is the tool capable of working the mining right in the Area over time, according to a scientific screen of sufficiency of technology for RDA.

4.3 In the Light of International Environmental Law

General international law set up two variables in the legal equation governing the Area: the first one is CHM, is declared in the binding manner by an act of the UNGA; and international liability for damages, for mistake or intent, in breach of contract.

There is no need to make a judgement about whether this discipline is good or bad: it has its merits and inconsistencies in the regulation of the Area (like other CHMs) and it was the possible regime to be admitted under a legal system of various sovereignties interests' whose interests are geopolitical views over the sea, which are often mutually exclusive (Koh 2016; Evensen 1986, pp. 483–498). Foregoing topics in this Chap. 3 explore these virtues and vicissitudes.

In turn, we can accept another perspective for approaching issues related to RDA in the Area resulting from mining activities: that of international environmental law.

Throughout this, the perspective is not tied to a formal and binding recognition of a certain category of assets, nor does it use concepts of breach of agreement in guilt. Emphasis from which the duties derive is the damage, the collective loss, potential or consolidated, to an asset (even if private or subject to a single sovereignty) of collective access.

The damage perspective changes the variables in the legal equation that will regulate the Area and fine-tune them with the perspective of international environmental law.

From the perspective of the legal subjects of the international environmental regime: the damage, potential or consolidated, in the Area, entitles international actors other than the National State or the international organisation, to the exercise of its powers: States and provinces "sub-national" municipalities and counties "local" and non-governmental—for-profit or non-profit.

Legal grounds for this exercise are much more harmonious with GPGs than any other institute that characterises the Area. While CHM is a concept worked on issues such as dominance and sovereignty, GPGs allow, from the perspective that damage to the environment does not respect borders, it engages in the concept of access and that precaution/prevention and polluter-pays are norms that any actor can hold to and/or demand—even if they are not at the international judicial level. GPGs operate at a level of success, *i.e.* that damage does not prevent multiple access.

GPGs operate at a level of governance[4] much greater than at a level of international responsibilities. That means to say the communion of international objectives, around the GPGs by voluntary acts of diverse actors. What actors are the ones? Whatever, since it tends to escape the process of coercion but of convincing: a group of non-governmental organisations (NGOs), sub-national or local government actors.

The lack of coercion processes leads to other consequences: an international legal regime based on GPGs, and its governance dispenses with the binding qualities of *hard law*.

4.3.1 GPGs, Sustainability, SDGs, and International Environmental Governance

In June 2019, the United States of America announced its intention to withdraw from the Paris Agreement as a result of the 21st session of the Conference of the Parties to the United Nations Framework Convention on Climate Change (2015 UNFCCC-COP 21), which was formalised in November 2019.

Nevertheless, two of its federated units[5] more populous, richer and with more delegates in the presidential election, they left in a just opposite sense, in addition to their largest city, and in a concrete way: the Republic of California, by ratification of the Paris Agreement 2015, assumes in its territory the commitments inherent to it, by institutional arrangement within the scope of its powers and State autonomy (CN544.2021.TREATIES-XXVII.7.d (Depositary Notification); the State of New York does so in a similar way through Climate Change and Climate Change Leadership Acts; as well as New York City, through Executive Order 26/2017.

Why does a State representative, which is not even national, challenge its central authority and make institutional arrangements to assume international commitments? In an international society whose subjects of rights and obligations are the National State and international organisations, how could sub-national and local governmental entities assume such commitments? And if they can, the force of this assumption?

Backing to the perspective of GPGs and its functionalities: what matters is not the balance of forces agreed on obligations and rights, but the damage caused to this

[4]On Governance, see Gonçalves and Costa (2011, passim).

[5]Other units have also taken similar policy action in the United States Climate Alliance initiative. See: https://www.usclimatealliance.org/state-climate-energy-policies. Accessed on: Jan. 23rd, 2021.

"compassionate" and, consequently, the collapse of the interests of the various actors (who are not subjects only) GPGs on the.

Therefore, the legal consequences that matter are not the "obligations" and the "command and control", "violation and sanction", "damage and compensation" regime, which are affected by the National State but the "naked and raw" of the consequences of the damage to the GPGs: these are going to suffer through all sorts of actors that not only the National State. People, producer groups, NGOs, cities, States, and provinces, will all suffer from the damage to GPGS and, in the light of this, everyone wants to act to avoid or face it.

Due to their inherent characteristics (seen in Chap. 1), GPGs are subject to individual decisions of collective/global impact. We understand this GPG as an environmental value/asset, just like the earth's climate.

In a neo-realistic scenario of international relations, it is assumed that the international actor, not only the National State, wisely operates their decisions in favour of the best result, or the optimum result, according to a value judgement of his own objectives (Slaughter 1998, pp. 21–54; Nye Jr. 2002, pp. 234–237).

In the case of a GPG, the likely analysis is that the individual behaviour—even of the strongest actor—uncoordinated from the others will produce an optimal result in the short term and terrible perspective in a medium—or long-term. The rational and selfishly fruitful—or optimal—choice for our actors is a common, collaborative action (Wolfrum 1999, pp.139–145).

Even in this neo-realist scenario, the interests of a given actor may not be consistent with those of the National State of which it is a subject. This implies in saying that this non-national actor, rationally, should plan and construct in the portion that has the power, to collectively ally themselves with the other actors whose result (optimal) as they see fit.[6]

It implies that the non-national actor will flee from the framework of the pure international legal regime, since their instruments are used to the basic principles of topics of international rights, in an institutional and organisational arrangement of powers. The exercise of power by a sub-national—for the collective management of GPGs will be much broader: would not only be regimes useful—as opposed to being the only means),—but all means of managing a common and global problem must be organised, oriented, and shared in a governance system. The regime turns into, in the words of Gonçalves and Costa (2011, p. 56), a species of the genre of governance.

Why does everyone? Number one, because not every actor in international society has the same type of authority or can assume the same type of obligation (Beyerlin and Marauhn 2011, pp. 245–256). See which sub-nationals and locals have part of (the energy resulting from[7]), as well as NGOs (i.e. public opinion and popular pressure, inspection, scientific data, legal know-how, capital) that allows the

[6]It is very peculiar how, depending on the point of view, the neorealism of global geopolitical analysis touches on constructivism, as both aim for collaboration based on a problem or global value. On constructivism as a current of international relations, see Slaughter (1998, pp. 45–50).

[7]Arising Constituent Power, see Moraes (2017, pp. 42–43).

management of a common and global problem must be organised, oriented and communal. All depends on the connection and the institutional and organisational links that this global governance will adopt (Gonçalves and Costa 2011, p. 56).

Briefly, governance is the system of rules that will coordinate efforts in the sphere of power of each actor in a position of getting involved in an issue and/or an objective, to maximise the results to the optimum. It is the norm of procedure, coordination, and cooperation around the common "good" objective.

Each actor category will contribute within its sphere of capabilities: *i.e.*, the scientific community and thematic NGOs exercising the condition of amicus curiae in Case 17 ITLOS (ITLOS 2011); Greenpeace, using its international capillarity, to embrace the 50-year moratorium request to the Antarctic Treaty, with the National States.

Note that a condition of an actor in international society elevated to the condition of subject of international rights imposes duties to this condition: cooperating is not an option for the international actor involved in the issue of GPGs. Anyone who has the right to, is obliged, according to the definitions of the principle of international cooperation (Aust 2010, p. 389).

Since the area is subsumed to the concept of GPGs we have already seen. It is important now to understand how the governance required by GPGs is manifested in the Area when it comes to environmental matters.

First one difficulty is at the local and sub-national level: due to the Area's own geographic reality (see Chap. 2), there are few points of touch among decision-making competencies and powers capable of influencing the exploration and exploitation discipline of this polygonal policy.

It does not mean to point out that local or sub-nationals are not able to act at all: it is only very limited due to other themes—such as the opening example of this topic, climate change). Considering the possibilities presented by the exploratory and mining procedure and within the local and sub-national competencies, there are still issues that pressure of a political nature with the government of the National State, corporate restrictions, fiscal parameters, and other measures can be worked with respect to the entities sponsored (contracted) to research or mining procedures.

Regarding RDA in the Area, sub-national- entities can act in governance informed by scientific criteria and aspects, *i.e.* science as a motivation for legislative and administrative normative acts (Mouton 1966, pp. 197–200 and 228–233), influenced by two axes of action.

The first one is, taking advantage of the dialogue between regimes of international environmental law, based on the very assertion of Case 17-ITLOS that the operations associated with research and mining of the international seabed that reach the water column are subject to the same responsibilities, then the portion of the area that is the object of the works (ITLOS 2011).

To consider this circumstance (*i.e.* pollution in the ocean water column), it is necessary to observe some similar cases:

> On 20 April 2010 the Mobile Offshore Drilling Unit (MODU) DEEPWATER HORIZON (hereinafter, the "DEEPWATER HORIZON" or the "Unit") was completing drilling operations at the Macondo well, Mississippi Canyon Block 252 OCS-G 32306 # 1, oil

exploration project in the Gulf of Mexico on the United States (US) Outer Continental Shelf (OCS) in preparation to temporarily abandon the well. During these operations, there was a loss of well control that resulted in a release of liquid and gaseous hydrocarbons, which culminated in explosions, fire, the loss of 11 lives, the eventual sinking and total loss of the DEEPWATER HORIZON, and the continuous release of hydrocarbons into the Gulf of Mexico. The flow was stopped on 15 July 2010 and the well declared sealed on 19 September 2010 (Marshall Islands 2011, i).

The example was not by chance: Marshall Islands is the Flag State and, as a signatory to UNCLOS III since 8/9/1991, had to open the investigation and take the measures provided for in the Convention and in the International Convention for the Safety of Life at Sea (SOLAS) concerning the maritime incident. Such conditions and obligations open the respective investigative report, with the irony of indicating the wrong date of UNCLOS III (1983) (Marshall Islands 2011, i).

However, despite the fact that the Marshall Islands is a State under the protection of and subject to the law of the United States, it is not a signatory to UNCLOS III; if it were, the accident would not have occurred in international waters.

And although the United States is not a signatory to UNCLOS III, a phrase sounded different in the Marshall Islands report, still in the editorial prolegomena: "Please submit questions to" and it follows an address in Reston, Virginia, with US telephone numbers and e-mail.

The fact remains that, although not responsible, the United States is the most affected State, and, as such, although not mentioned in the UNCLOS I rules, it was compelled to collaborate in the investigation. In accordance with the principle of sovereignty, the USA could not even be considered a Coastal State or empowered with this condition, since it is not a signatory to UNCLOS III.

But nonetheless, and in the most evident collaborative exercise—regardless of motivation),—the USA produced their 244-page report on the same fact, in coordination with the flag country. In this American report, another curious circumstance arouse: in the American tradition of attaching all the secrets and/or stamps of the entities that participated in an operation, amid stamps of agencies—such as NASA, EPA or OSHA—government departments "Wildlife and others", forces (Coast Guard) and specific task force for the case, some seals call our attention in light of the subject matter of this topic: the seals of the States of Alabama, Florida, Louisiana, Texas, and Mississippi, and the seal of an NGO (the Coast Guard Reserve).

Fact is that although sub-national and locals are not exactly "Coastal States" (articles 7 and 69, UNCLOS III), and in our case neither the US is, and, as such, entitled and entitled to postulate about their own coastal interests in view of the Convention and the environmental disaster, the management of this disaster would not be the same without the collaboration and governance of these players.

How important it was in such a way that the USA, although not responsible for UNCLOS III, became a focal point for dialogue on both reports.

It should be noted that the works related to the accountability of the individual and the Flag State were, to the extent of the powers, conducted by subnational-entities.

Ironically, there was not a word in both reports about the Area. But the example consolidated our point of view: in the face of unique needs—Marshall Islands should avoid greater accountability and the USA avoid an even worse tide and disaster—nations and sub-national- entities have joined forces.

What about science? Across the two reports, each step and decision were guided according to aerospace and geographically consulting (National Aeronautics and Space Administration (NASA)), geological and geotechnical (United States Geological Survey (USGS)), and environmental protection and recovery (Environmental Protection Agency (EPA), Wildlife, Parks) (Marshall Islands 2011, passim).

The second is, it is about the range of powers, the ability of a subnational- to act in a case for which the rules of regulation refer to a subject of Sovereign Nation State rights, not autonomous or local.

Only a few things or even nothing can be done. However, Wolfrum (1999), due to the transversality of the systems in view of the infrastructures that support them thinks the opposite: sub-national and local governments and NGOs have their own range of powers: economic, fiscal, or even normative.

As an example, advised by science, the use of certain technologies that are less aggressive to the environment for the disposal of oil production (in the high seas—or even their replacement—can be stimulated in subnational—territory through tax incentives of a tax nature. So did the Port of Los Angeles, for example, to take on the International Maritime Organization (IMO) 2020 commitment for lower sulphur bunker fuel: long before the IMO rule was in place, the Los Angeles Port Authority did not allow docking of ships in disagreement with the rule, by a disciplinary deliberation within the port.

It allows us to notice, then, the legal norm that, par excellence, intends to condense the governance mechanisms that will give GPGs the necessary reach.

As an integrating norm for the governance of the funds, we can proceed to a critical analysis of what was considered the legal norm related to the new global integrating systems par excellence: the 2030 Sustainable Development Goals (SDGs).

Approved on Res. 70/1, UNGA of 12/25/2015, its content is non-binding, the SDGs deal with 17 objectives expressed in 169 target objects of action by international actors. They are called actors because any international actor (States, sub-nationals, locals, international organisations, NGOs) can adopt them as a goal to their scope of work, as already addressed in this topic. The main goal is to focus on SDG 14, whose object is the sea.

This section deals with a critical analysis of the SDGs, centred on SDG 14 and its medium expressions written in Objective 14.c. The analysis confronts the reason for the existence of the SDGs, *i.e.* their goal and means, their function as an international legal norm, with the precepts it carries or references in 14.c.

To understand the sea as an international legal object of interest to the SDGs, especially SDG 14, it is mandatory to keep in mind the following:

(a) The sea, like the dry surfaces, is a mosaic of biomes and ecosystems that extend along the horizontal axis of the landscape and, unlike the dry, vertical surface of the water layer
(b) Being that the largest of the earth's crust biomes, contained in the marine super-biome, is still the minimum studied and explored.

Based on the premises, we must seek, in critical analysis, the effectiveness of the legal paradigm created by SDG 14 (14.c), in which it declares the UNCLOS system as "the legal framework for the conservation and sustainable use of the oceans and their resources", which was confirmed §11, of the Declaration "Our Ocean, Our Future: Call to Action", final document of the United Nations Conference on Oceans, June 2017.

From this relationship, we ask if the precept of SDG 14 (14.c) would not sabotage the reason for its own existence: a new format of precept and international normative effect anchoring its contents in formats of the 1970s and 1980s, codified and stratified, with effects declared ineffective before achieving and advancing sustainability.

For this goal, we sought, from the perspective of the normative precept, the origin and function of the SDGs and the 1982 United Nations Convention on the Law of the Sea and their consistency in terms of the effectiveness of a principle, namely, that of sustainability applied to uses of the sea and the Area. The investigation was based on (a) both the usefulness of the international normative precept on the phenomenon in question as in (b) sufficiency of the actors/subjects of law involved.

In the end, we are planning to answer the questions: Are SDGs necessary for the international legal regime of the sea? Do SDG 14 and 14.c help or prevent in achieving and advancing this principle of international law?

It allows us to take the perspective of the need to solve—or, at least, face—major global problems—environmental, among them and mainly—and the honesty in saying: the international regimes built so far have mostly failed. The complete proof of this failure is in the evolution of the international legal language: if in the 1970s and 1980s, of the Section XX, the "Age of Codifications"[8] addressed major environmental problems, even if in a fragmentary and thematic way, in a preventive and cautious character in face of the risks that are created by them, today the speech is of contingency of risks already realised, materialised (Beyerlin and Marauhn 2011, pp. 441–444).

Thus, to understand the sustainability goals, we must bear in mind that this is not a new normative format that was necessary to face the global environmental problem but an innovation in the use of the old preceptive format. It allows us to know the whys.

[8] The Age of Codifications is the period of International Law and the Law of the Sea, concentrated in six decades (1930 to 1990, of the twentieth century), in which the International Society produced (especially for issues related to the Law of the Sea) treaties markedly marked by preceptive constructions in "coded" modelling, comprising rigid, binding, watertight structures with the intention of enduring (Scovazzi 2000).

Denny and Granziera (2017, pp. 14–19 and 20–24) point out that the "Tragedy of the Commons"[9] goes into its new act, being as current today as the time in which it was written: the problems had erroneous approaches and the language of manifestation of competences over it that were even more inadequate.

Carlarne (2014, pp. 8–9), when discussing climate change, from the perspective of the effectiveness of solutions to the problem, points out the approaches mentioned by Denny and Granziera (2017, pp. 20–24). It allows us to take a look at the following perspective.

The International Society was struggling with the global problem and lacked actions that would make the management of the common environmental heritage effective. (a) The International Society insisted on restrictive interpretations, due to its hierarchy and its subjects of rights, in derivation to interpretations conservative sovereignties derived from the Münster and Östnabruck-1648 Treaties (Abi-Saab 1996, pp. 51–53; Schrijver 2008, pp. 379–380), much in function of the crisis itself—in the 1980s and 1990s—that the State's legal-political institute (Giddens 2011, pp. 450–452); and (b) pointed out normative needs of a preceptual nature regarding a supposed lack of *gründnorm* to support, articulate and integrate international law in search of an environmental result (Abi-Saab 1996, pp. 46 and 122; Carlarne 2014, p. 29).

Environmental issues had led national states into focus and the vicissitudes that have led them to the contesting crisis of their utilities and functions: too big for small problems, too small for global problems (Giddens 2000, p. 23). This, however, does not require an attitude of abolition of the subjects of the sentence! According to the exposition, we can think from the perspective of Carlarne (2014, pp. 33–41), that of the object of the sentence: the problems, and different problems require different approaches.

In other words, do global environmental problems, in their various facets, require a *gründnorm*? The author discloses that environmental problems, due to their peculiarity of global risk management of multiple unequal actors, do not require a *gründnorm* and sometimes it gets in the way (Carlarne 2014, pp. 42-45).

The environment, since it is a problem that has an endless polynomial of natural variables as a result and consequence,[10] it needs solutions that express an equally

[9] Although the concept of Commons finds correspondence in legally old institutes (such as the compassionate one), it was Hardin who characterised the environment, globally, as a Commons par excellence and pointed out, by criterion and game theory, the "Tragedy" that free Common's administration. Commons, in this case, is the common heritage, of collective use and of responsibility for collective maintenance. In this sense, see Denny and Granziera (2017, pp. 9–27).

[10] UNCLOS Article 136: "The Area and its resources are the common heritage of mankind." UNCLOS Article 137.2: "All rights in the resources of the Area are vested in mankind, on whose behalf the Authority shall act. These resources are not subject to alienation. The minerals recovered from the Area, however, may only be alienated in accordance with this Part and the rules, regulations and procedures of the Authority."

UNCLOS Article 145: "Necessary measures shall be taken in accordance with this Convention with respect to activities in the Area to ensure effective protection for the marine environment from harmful effects which may arise from such activities. To this end the Authority shall adopt polynomial approach and that consider these endless variables in syntax. Thus, one wonders how a *gründnorm*, be it a nineteenth-century framework convention, whether it is a seventies/eighties codification, can it fulfil this objective?

"Framework" norms and codified standards are unable to fulfil the very peculiar need for the approach of the global environmental object. Not, at least, in isolation. Carlarne (2014, pp. 19–23) and Beyerlin and Marauhn (2011, pp. 441–446) state relevant reasons:

> The international legal norm must allow a power of action, of force, based on the collective, and not on individual pillars of power.
>
> The norm tends to form cooperation mechanisms, and to move away from confrontation mechanisms.
>
> The rule tends to seek preventive and aggregating controls instead of repressive and disruptive ones.
>
> The standard must seek compliance and its assistance, when it fails or when it points out non-conformities, instead of sanction mechanisms.

It allows us to apply the concepts to the sea and the Area.

Coding is the format of the current UNCLOS III. And as such, it has not neglected to address global environmental issues. However, we observed that such an approach, as evidenced by the perspective of the effectiveness of solutions to global environmental problems, is notoriously insufficient (albeit courageous).

An environmental system the size of seven seas, including the largest biome in the world in extension and the least known (*i.e.* the seabed) (Groombridge and Jenkins 2002, p. 119; Taylor and Roterman 2017, pp. 4872–4875; Datta 2011, p. 189), global environmental problems are also treated in a coded way, in thematic compartmentalisation and under a temporally dated and obsolete prevention perspective. Scovazzi (2000, pp. 88–89, 122–124, 206–219) enunciates environmental topics addressed and the cycloptic form enunciated, still fragmented, including gaps to be answered, and, in a dreadful tone, points out that the international seabed, although they are CHM, they are almost exclusively regulated environmentally, in effect, for mining purposes. Menezes (2015, pp. 177–188), on the other hand, does not deny that there was an evolution through UNCLOS III, which increased a binding nucleus of environmental object norms (incorporating Dec. do Rio-92 in the environment regime the seabed surrounding area), but it does not yet deny the prevalence of soft law in internal State cooperation criteria, in identical regulation prior to the SDGs. Schrijver (2008, pp. 257–258), in turn, in analysis of the principle of sustainability in international law, points out that UNCLOS III is the product of the North-South geopolitical confrontation and, as such, was timid for the

---

appropriate rules, regulations and procedures for inter alia: (a) The prevention, reduction, and control of pollution and other hazards to the marine environment, including the coastline, and of interference with the ecological balance of the marine environment, particular attention being paid to the need for protection from harmful effects of such activities as drilling, dredging, excavation, disposal of waste, construction and operation or maintenance of installations, pipelines and other devices related to such activities.

advancement of the matter, using standards of contained or limited effectiveness, with the need for specific topic technical provisions, in a controversial discussion environment.

And, even among these regulated segments, there was no exemption from disasters and deregulations in environmental matters. Note that oil pollution of the sea, even after the *International Convention for the Prevention of Pollution from Ships/*2013) or the Lessons Learned of the International Maritime Organization (IMO) were little incorporated in sequential regulations; and some environmental damage resulting from accidents or mistreatment with the common environmental heritage could only be identified today, through the exploration—investigation—of the seabed (Foden et al. 2009, pp. 22–24), although legally, the regulations have already exhausted the field of their respective efficacies.

It is evident that, in view of what has been observed, there is a mismatch between speech and practice: it is not the lack of *gründnorm*. Where it occurred, there has also been no progress in precaution, prevention, control, contingency, or accountability for the mistreatment of global and common environmental heritage.

It is evident, from this perspective, the need for an innovative approach to the global environmental problem at the level: a new structure of international normative precept that allows the approach of environmental problems to accommodate all the complexity that it has and represents.

We talked about the perspective of the actors of the global environmental problem.

Not only from the perspective of the object that the precept and international legal norm require new formats. It allows us to dwell again in the phrase of the twilight of the Section XX: the environmental issues brought national states into focus and the vicissitudes that led them to the contesting crisis of their utilities and functions: too big for small problems, too small for global problems. If we see the problem from the perspective of the object, it can also be taken from the perspective of the subjects: the actors of the International Society.

The International Society, despite being perceived as a formal legal construct—although doctrinal—does not escape phenomenology: it is a phenomenon that "is" and not that "should be". The International Society, as such, is a dialectical stage for international actors that have always existed. What varied in time and space were the vectors of the dialectic between them: in intensity, direction, and sense of power that they concentrated.

In the case before Section XV National States were not "subjects of rights in international law", this does not mean that they did not exist. They existed in various forms but were necessarily less powerful and their relevance to the construction and evolution of international law was relegated to other, smaller—or less demanding power—roles in facing international problems. Treaties and wars were fought, agreements and trades between peoples and nations were signed, and crowns and sovereigns were either recognised or not by their peers. International law existed and was a real phenomenon (Abi-Saab 1996, pp. 45–46; Douhan 2019, pp. 1–2), but it manifested itself in different ways than the international law practised today.

It was only from the rise of the State as a leading figure in the national and international dialectical theatre of operations—which happened, as a general Western phenomenon, in the twilight of the fifteenth century),—that the Treaties of Münster and Östnabruck-1648 aimed at the phenomenon "Sovereign State" passed making sense as a phenomenon in international law, ushering in an era in which the concept of sovereignty becomes its principle of choice (Abi-Saab 1996, pp. 46–49).

Today, the International Society, as a rule, considers as a subject of rights only National States—that is, sovereigns—and those entities that, due to the sovereign will of National States, can assume rights and obligations in their own name—such as international societies).

Although sovereignty is—by historical analysis—considered a norm-principle, this principle has hesitant optimisation episodes to address the "wicked problem" global environmental impact. It is because of the following reasons:

(a) Highly figure of the State began to be questioned in the 1980s (of the twentieth century), in relation to its practical utility to society (Giddens 2000, p. 23) in contrast to the amount of power it focuses on which led the State to enacts itself in defensive positions of its own position, putting less effort into other variables—let alone globally common environmental problems, which would require flexibility of concepts of sovereignty, by definition, concept and origin)
(b) Global geopolitics also has the dialectical field of capital production as a function of its results, *i.e.* global trade, industry, and finance. This perspective still addresses the environment as a resource. And, in a competitive environment, giving up resources is giving up competitiveness. Which leads sovereign National States to the "prisoner's dilemma" at the international level (Gonçalves and Costa 2011, pp. 134–139) and to conflict with individual short-term objectives with common long-term objectives (Beyerlin and Marauhn 2011, p. 440).

Norm-principle, according to Dworkin (1982, p. 42), is defined as follows:

> J'appelle «principe» a standard that does not exist, not a partner that is a realist situation in a political, social or social situation, desirable, but a partner that is a requirement of justice or of l'équité or bien d'une autre dimension de la morale. Also, the standard that is prescribed for the automobile accident is a politique, whereas the standard that does not provide protection for the commis is a prince. On peut effacer cette dis- for example le principe selon lequel que cette politique implique un but valable) ou encore en adopte la thèse utilitariste selon laquelle les principes de la justice sont des affirmations grand nombre). In certain contexts, the distinction is a utility, which loses itself on the face of cette manière.

Rittel and Webber define the expression "wicked problem" as that phenomenon that gathers ten peculiar characteristics, although they have done it in four topics: (i) the problem does not have a single and definitive formulation and depends on the subjective perception of the actor who observes it (and eventually addresses it); (ii) its criterion of totality and its complexity do not allow the description of the problem in a temporal or spatial character from the beginning and end; (iii) the variables are multiple and not binary, that is, the criteria of true or false of the variables of the problem are replaced by solution gradients that depend on the reference framework of the problem (such as a notable product in a polynomial in an environment of theory of Gaussian sets); (iv) there is no possibility of testing or testing the solution; (v) given the impossibility of testing, the action is taken and generates immediate consequences for the problematic phenomenon, consequences that must be considered for the next action on the problem, since indelible; (vi) solutions do not have a single response process; (vii) each problem and its component variables is unique; (viii) each problem and each variable is the cause and/or consequence of another problem and its variables, with a relationship like the nodes of a spider web: moving a knot or its thread reverberates in another knot or chain thread; (ix) the language explains in many ways (which can conflict or confuse) the problem; (x). The problem has direct syntax with the rights and well-being of many (Rittel and Webber 1973, pp. 155–169).

This does not mean that other international actors (prior to the 1648 milestone) disappeared or that some did not appear. In fact, in the face of this wicked *problem*, in which the State demonstrated its political and legal ties to act appropriately, the other international actors were able to show usefulness and value. Rei (2017, pp. 33–36) points out:

a) the discussion or forwarding of global environmental solutions, when directed to results, is always originating, or backed by scientific values pointed out not by the States, but by other actors that lack Sovereignty.
b) solutions go through geographic and political levels that are not necessarily the National State and its political ties, taking the shape and strength of subnational (state and provincial) and local governments, whose political ties are different, which represent less impediment to the achievement of conducting the tasks. solutions in their geographic space. There are needs for new actors, not tied to the same competitive or governance "dilemmas" that sovereign National States were in.

Therefore, the modelling of an international environmental legal standard codified or based on production processes linked to a particular modelling or methodology was not suitable for addressing the environmental problem and its multifaceted aspects: one, because the State was too tied to extra-environmental issues to seek the appropriate consensus; then, because the solution or implementation of solutions to environmental problems required the participation of other actors (sub- national—and local governments, scientific and social actors and subjects—who, in turn, did not participate in the construction of the normative precept.

Again, it allows us to address sustainability in the general regime of Law of the Sea until the moment built. In general terms, Tanaka (2018, p. 112) enunciates some keystones of sustainability in application that were already enunciated in this work.

At this point, and for now on, there is harmony between the statement in SDG 14 and 14.c.

In the same essay, about the nature of sustainability and its effectiveness in international law, Tanaka (2018, p. 131) still explores, through the analysis of precedents of international courts:

(A) the T nature of sustainable development in being an integral key concept, of mandatory observance to judge the conduct of the international actors in dispute about the way they managed/managed the use of the natural resources at their disposal and/or to be an integral key concept, mandatory compliance to judge the conflict between international standards; and b) the principle norm derives from an interpretative process that is fundamentally temporal, evolutionary, in the terms informed by article 31 (3) (c), of the Vienna Convention-I/69.

Applying the concept of Dworkin to the species of Tanaka, sustainability is, among other things, the norm of the (international) Law of the Sea. The interpretive key, according to the author and considering Rei's words (2017, pp. 33–36) will be, fundamentally (to say the least), science.

(B) Sustainability in treaties, when their effectiveness is controlled by resolutions, tends to be interpreted as a binding rule, within the limits of the treaty, even if it is not.

Such a position which leads us to advance in two assertions: that non-binding norms are not so by "binary" criterion—it is or is not *binding*),—but by a criterion—as previously observed in general terms in this essay—verification gradient, the medium of which is legal hermeneutics and the criterion is science as well as the need it indicates by adopting the or not sustainability in an activity regulated by treaties.

Not without reason, and in the same sense, Case 17 ITLOS structured the binding adoption of principles of the Declaration of Rio-92—in pure and isolated non-binding analysis—by the UNCLOS States Parties, regarding the adoption of measures internal laws regarding the accountability, command, control, and compliance of individuals sponsored in the seabed mining.

(C) Finally at last, the possibility of sustainability being the cause of requesting any rights subject named in a treaty to defend the interests and objectives of this treaty, even if this subject is not affected—not, at least, directly—, by the actions or omissions of those who opposed sustainability.

Many different languages address the same problem: now sustainability (here taken as an offshoot of GPGs), now CHM. An attack on sustainability means an attack on a *common*, to the "compassionate" that the International Society takes advantage of, either directly, for the direct damages and future scarcity caused, or indirectly, for the attack on the sustainability value—as a positive legal norm or as a resource management dynamic—, or due to the complexity of the problem that, one day, it will evolve and reach its own interest. Cançado Trindade (2006, pp. 365–367) points out that CHM is a non-exclusive principle of the Law of the Sea and consists of two generally applicable clear rules:

(a) A wave, coming from the word *Heritage*, in the sense that it is about the sharing of a good or value in society thought of in time, *i.e.*, generations to come should have their share of the benefit of this good and the obligation to preserve this good in an equal way to their predecessors, in everything approximate—the author insinuating coincidence—to the principle concept of intergenerational equity[11]—intergenerational equity, in the original—, for this portion of the CHM; and

(b) Another spatial/geographical one, derived from the word *Common*, that all international actors—not necessarily subject to international law—are responsible for the conservation, for the inspection of this conservation and for the extraction of benefits from these goods or values.

Costa de Souza (2019, pp. 51–61) comes to the same conclusion at the administrative and economic level (infrastructure, par excellence): with the aim of the Law of the Sea, especially the seabed, in the absence of a more developed concept CHM was used with purpose and effects almost coincident with those of sustainability—if we compare the definition of Cançado Trindade about CHM and Tanaka about sustainability. The "Our Common Future" Report, which contains the (scientific) definition of sustainability, which would serve as an integrating and interpretive element of the principle of sustainability in the production of international legal standards, was only published in 1987, 5 years after the signature of Montego Bay/82 and the speech by Arvid Pardo, on 1 November 1967.

So, as such, it is an attack, at least in a finalistic character, on the sovereignty of the counterpart of a treaty itself: if it is positive and it is agreed to take geopolitical steps towards the sustainability principle, doing the opposite means an attack, finally, the recognition that is given to the sovereign wealth—in the sense of the collection of rights and obligations—of the counterparties. All counterparties.

The needs imposed by sustainability, highlighted above, make it necessary to recognise that the State is insufficient to complete, alone, the very precept chosen by them in a treaty. There is a need for a greater plethora of international actors, in preventive action and cooperative advice on environmental compliance, which increase the hermeneutics of precepts loaded with sustainable concepts.

UNCLOS III was no different: the combination of its Part XI—mining in the Area, name of the international seabed cf. article 1 (1)—with Part XII—environment and preservation of marine biomes—clearly highlighted the points addressed by Tanaka and Beyerlin and Marahun: technical insufficiency of the precepts, which imposes methodological procedures for command and control of processes by unitary administrative bodies; the need for scientific reading of the precepts (historically considered), which requires the aggregation of other international actors, in recognition of the failure of States to comply with precepts that they carefully chose for themselves.

In this sense, we have Case 17 ITLOS as a curial to observation for confirmation of the essays by Tanaka, Cançado Trindade, and Beyerlin and Marahun.

The International Seabed Authority (ISA) is not exempt from provocation. In fact, UNCLOS States Parties necessarily seek the Area Authority not only for their

[11]On the Principle of Intergenerational Equity in International Law see WEISS, Edith Brown. Intergenerational Equity: A legal framework for global environmental change (Weiss 1992, pp. 385–412).

mining authoriation regimes, but also through consultancy, which generated Case 17 ITLOS.

In short, ITLOS Case 17 interprets the obligations of an environmental nature in the case of mining in the Area, under the terms of article 139, UNCLOS. We can split them into three distinct categories, but right now, observing the duties of the States Parties and those excluding liability and unlawfulness, excluding modalities of objective accountability:

1. The precautionary and preventive obligations of the Environmental Impact Study of the mining activity
2. The preventive obligation of due diligence and the establishment of environmental quality standards in the execution of installation projects and mining operations
3. The State Party's obligations and exclusion of its liability in the event of damage and how to respond to this damage (indemnity and/or reparation)

There is no mention, either in Part XI or in ITLOS Case 17, of any post-operation environmental liability or obligation other than those arising from an unlawful civil/contractual or international administrative act. In other words, there being an environmental obligation after the mining operation, there must be two verifiable elements: damage and unlawful act—attributable to the State Party for action or omission—supervisory or normative—(ITLOS 2011, pp. 206–211).

If the interpretation of mining activity in the Area were restricted to Part XI, UNCLOS III, and ISA regulations, it would be understood that the State Party that established a sufficient environmental regulatory universe and endeavoured to inspect the mining company would be free from any environmental liability.

However, sustainability has a double helix of interpretation and implementation: regarding the seabed, economic development will be in line with the legal category of CHM.

And it is at this point that the legacy of ITLOS Chief Justice Tullio Treves goes far beyond the first unanimous vote of the Court: to make the dialogue between Parts XI and XII of UNCLOS and carry a CHM close to sustainability, with meaning and effectiveness, brought to the I cut postulates by *amici curiae* that are not classically considered subjects of law, but, rather, NGOs and international agencies, and have also made available reports—but out of the file for extemporaneous—of other NGOs. ITLOS, at least this small point, makes the sea regime bow to the new needs of international law to achieve this particularly important common objective, sustainability.

Nonetheless, even if we are talking about a successful advance in the incorporation of other international actors and expansion of the interpretative scope of the regimes, integrating them with common principles expressed "even" in non-binding norms, it is good to put things in a realistic perspective and conservative: it is an organ, of temporary members, that issued an advisory judgement for a specific item of a specific treaty (although the main one) of the international regime of the sea. And, even so, the decision is not free from the vicissitudes resulting from the insufficiency of the precepts and the actors mentioned in 2.1. and 2.2. of this one.

Egede (2018, p. 158) highlights: although the States Parties and their authorised agency, *i.e.* ISA, have the responsibility to regulate all international seabed activities and their resources in general (UNCLOS, articles 136, 137, and 157), so far there has only been normative regulation of mining in the Area. In other words: the governance of environmental objects of global interest at the expense of a single agent, in process regulation models, in administrative and police command and control methodology continues, for 25 years, to fail by default, depending on other regimes (fragmenting the condition of UNCLOS III as a single coded body to be interpreted), such as the Framework Convention on Biodiversity to integrate commands that are still lacking (Scovazzi 2000, pp. 213–219).

This finding, *per se*, already leads to confirmation of the assertions of Carlarne and Bayerlin and Marahun of the need to evolve the normative models of international environmental law, which could be met by the SDGs. However, the questions do not stop there: just as in the form of the precept (or in the absence of it, as Egede points), in the content there are also problems in relation to the Sponsorship States and the ITLOS Case 17.

As a way to defend the environment of the area from disasters in exploitative mining, Case 17 ITLOS imposed various interpretative provisions of Parts XI and XII, of UNCLOS. Among them, it rejected the application of the principle of common but differentiated responsibilities (CBDR) the liability of the Sponsoring States for environmental damage in the Area resulting from their misconduct. CBDR was invoked and regulated for the purpose of Case 17 ITLOS in paragraphs 131 and 151 to 169 of the judgement reports (ITLOS 2011, pp. 45, 52–57), during the analysis of the taxable event, verification of the fact generator and the limits of the liability of States Parties for damages in the Area, and not of the entities sponsored by them for the execution of mining in the Area.

Brief digression about CBDR: when the legal norm tends to treat non-uniform subjects of law uniformly, some initial phenomena are noticed: ethical and practical.

The first of an ethical nature, deals with a criterion of justice: giving the same legal treatment to those who do not have the same capabilities does not mean equality, but privilege to inequality, which turns out to be an injustice. This concept is not new, neither in the superstructure nor in the infrastructure of phenomenological thought.

The second of a material nature, means that in the face of a need to affirm an international norm, the subjects of international law who are more fragile (or less powerful about the object of the norm) are naturally not inclined to assume it unless their treatment serves the rebalancing of forces between parties. Thus, a process of different treatment, compensatory in powers, rights, and obligations, would have the ability to attract more parties to the international standard, which would give it more scope and legitimacy.

The third, also of a material order, but infrastructural: the CBDR must be described as consecration of the idea that the mode of production and consumption of the Section XX—post-World War II can no longer be the standard between peoples and nations, under penalty of aggravating global environmental problems and generating others (Beyerlin and Marauhn 2011, p. 64).

However, the trial did not remove it from other concepts and phases of mining in the Area, reserving differentiated discipline to "developing states".

They, and UNCLOS, failed, however, to define "under development". Such failure, in addition to the obvious condition of extending the North-South geopolitical conflict (see COP Copenhagen—UNFCCC), which is absolutely undesirable, generates legal inaccuracies in the face of an activity that needs legal certainty in the face of its problems, operational costs and inherent environmental risks (Egede 2018, pp. 159–164).

Still, Egede (2018, pp. 169–171), even if it points out to the Nauru/Tonga model of Sponsor State differing from the "Sponsorship-of-Convenience" – in allusion and direct analogy to the navigation convenience flags and their inherent disasters, based on article 148, UNCLOS, do not escape the grey areas and insufficiency acknowledgments:

a) It was necessary to disregard vertical sovereignty to act on a private international actor for the purpose of accountability.
b) ITLOS did not define a closed scope of liability and its incidence hypotheses, nor did it often define the content/scope of the Sponsoring State's legal and liberating due diligence. What about this due diligence? Is it a duty of the Sponsoring State, or is it the prerogative/option of the Sponsoring State to assume the responsibility of the sponsored mining corporation? The structure with which it was built the Case 17 ITLOS Report intends to answer the second hypothesis.

Although UNCLOS III is the most advanced international coded, it is insufficient in the face of a global terrible problem for the same practical and political reasons that made up it. UNCLOS III is not a bad standard for sustainable development: at the time it was written and according to its possibilities, it was a great victory, including the opening and regulation of new markets and sustainable development. But then, as said before, they had no tools of their own to go further.

Since either by the object or by the subjects, there was a need for a new normative modelling that would allow the participation of new actors, the transit of relevant scientific information, and a methodology for addressing a problem of complex syntax variables, law international addresses an old normative form but in an innovative way: it resurrects the use of soft law, burying the "Era of Codifications". Here are the reasons.

While it is not binding, soft law models behaviour. Beyerlin and Marauhn (2011, pp. 33–34 and 37–38) point out that these norms have the power to generate an effect of directing the behaviour of the actors of the society in question: their bias, to a greater or lesser extent, their acts towards to approach the key concepts of soft law or the values it advocates and curb the taking of actions or omissions in the sense of ignoring these values from the computation of their global positions.

Although its precept is authored by sovereign National States, the construction of its respective legal norm—behaviour modeller—is open to the dialectic of all the actors of the International Society in real time, building and structuring it in terms that they understand to be adequate, forming a historical-geographical synthesis of interest. Is this concept applicable to the formulation of international legal norms, especially those dealing with global problems? Based on the conclusions of the analyses by Beyerlin and Marauhn (2011, p. 256), the answer is in the affirmative, especially in the case of international environmental law.

Bear in mind that in all legal norms, the construction of its fundamental elements of production—precept and interpretation of the Court, the administration, the trustee—are subject to influence by information brought by agents external to the production of the legal norm. It was these agents that identified the nominally producing actors "of norm and precept"): (a) the problem; (b) the data and hypotheses for solving the problem; and (c) Dworkin's current "ethical paradigm"—in the moment the lack of better expression—to be applied to the precept.

In addition, the fact that those external agents are much more than subnational- or local governments (to use the same nomenclature given by the United Nations. In their (natural) majority they are non-governmental organisations (NGOs), subnational- and local governments, defined as entities that they participate and manifest themselves in international society as forces of pressure, and, although they have legal personality under domestic law, international law grants them a certain legal status, which in turn will vary according to the dialectical arena in which the entity intends to influence (Beyerlin and Marauhn 2011, pp. 256–257), but hardly the condition of subject (of some) right.

Although it addresses the same (*terrible*) environmental problem, soft law does not impose criteria and techniques of production processes, inspected by police command and control measures—typical of power administrative police, even. Its operating regime is based on the establishment of objectives common to all National States and other actors, informed by objective technical-scientific criteria, and intermediate checkpoints. The process for reaching checkpoints and the final goal is free. What makes this type of norm more amenable to adherence by different actors from diverse cultures and fulfilment of these objectives more friendly, since the local culture will not need to denature in a technical process imposed on production. In turn, the form of control will take place through the governance of the different stakeholders, increasing transparency, discussion, and the capacity of the actors to mobilise in the face of the problem: subnational- and locals that adopt the criteria that their respective national ignores, forcing the behaviour of this and its peers (*i.e.* São Paulo's adoption of climate change goals before the Federal Union did so); NGOs that disclose technical media liable of extra-legal disapprovals (*i.e.* case "Whaling in the Antarctic" by the International Court of Justice and Case 17 ITLOS); adoption of commands and controls between regimes as a way to optimise the progress of the achievement of the SDGs (*i.e.* IMO 2020—a regulation within the scope of SDG 14, which aimed at SDG 13).

Based on these concepts, the "SDG" points to the International Society objectives, checkpoints, and ways of controlling 17 thematic action groups, with the objective of sustainable development as scientifically informed in the 1987 Our Common Future Report.

The sea was contemplated with an SDG 14 all its own, being one of the two geographical spaces to deserve a peculiar SDG (the other was SDG 11—Sustainable

Cities). SDG 14 deals with "[c] preserving and promoting the sustainable use of oceans, seas and marine resources for sustainable development". It has seven partial goals (MDGs 14.1 to 14.7) and three methodologies for achieving them (14.a to 14. c).

The importance of this SDG 14 transcends oceanic sustainability in its uses: its importance operates on two additional levels, namely, the syntax with other SDGs and the syntax with the geopolitical balance of global power on its maritime boundary.

In the first syntax, there is the sea as the geographic site of 96.5% of the world's water, as well as being responsible for 50% to 85% of the atmospheric O2 production, as a reserve and sink for 25% to 31% of the CO2 released on the planet, as a food reserve of 200 million pounds (of protein and dietary fibre) for the diet of the planet's population, as the planet's climate regulator through its oceanic and atmospheric currents, and as the medium through which more than 90% of the cargo traded worldwide passes (NOAA 2020). The sea is much more than this (see the religious meanings that diverse cultures attribute to the sea), but just from these examples given in this paragraph we can categorically say that there is no SDG fulfilled in disassociation or disarticulation with SDG 14, with the sea and its eventual sustainable use.

The second syntax deals, precisely, with the dominance and geopolitical struggle between sovereign nations in their world order of forces. The ocean has always been the scene of international disputes, having, almost all the time, the sea served as an instrument of geopolitical power for dominance of the global force scenario (Barbosa Júnior 2012, pp. 209–219). The examples are lavish: since the emerging of international law (still in its liberal Enlightenment character), the clashes over the dominance of the sea between supporters of the *mare liberum* (France, England, Holland, and Sweden) against the supporters of *mare clausum* (Spain, Portugal, and the Pope), and there were still those hybrids in position (the Duchies of the Italian Peninsula, Russia, and Denmark). And they keep on with today's dispositions and conflicts at sea and their dominance by military power (Scovazzi 2000, pp. 53–116). Nobody can be sovereign in the seas due to the express disposition of the international news on the sea since Geneva/1958, repeated by UNCLOS III (positifying and objectifying what was jurisprudential and customary). Which means that whoever can apply power in this geographic location will make a difference, even though UNCLOS calls nations to cooperate in activities on high seas (see articles 117 to 119 in the field of fishing on the high seas), that was seen as insufficient, the view of the fishing wars in progress (Scovazzi 2000, p. 132).

This second syntax is a variable that impedes the action of National States—since it is more a complex problem to add to a terrible problem),—which would risk (at least in theory) the achievement of (as seen) all SDGs. However, due to their open governance and execution format, subnational- and local (and even private) entities can assume these commitments within their operations and geographic regions, minimising geopolitical risk.

So if the SDG system is that good, what could have gone wrong? Why can't the shot backfire?

The geopolitical syntax of power and dominance that the sea has meant over time, has already been explained at the end of the previous topic, it was one of the main variables in the evolution of the Law of the Sea. In turn, the intricate evolution of the Law of the Sea in international law did not take place on its own and was not intricate for nothing. Until we reached to the different regimes of law in the different geographic zones of the sea (*i.e.* territorial sea, the Area, international waters, exclusive economic zone, contiguous zone, continental shelf) and what each regime links (normatively) to a dialectic between nations, went through more than 530 years of wars, international judgments, doctrinal postulations and, even, normative tests by trial and error (Scovazzi 2000, pp. 53–116).

It is worth mentioning that the current UNCLOS III, although it does not have the formal assumption of the treaty by the United States of America, brings together the international consensus on the discipline of the immense and significant international dialectical arena that the sea means and with obedience (at least in part) of the United States, since UNCLOS III consolidates aspects of UNCLOS I and customary Law of the Sea.

UNCLOS III oversees a product of international coding for preparation and foreseeing since 1967, the result of negotiating rounds that only ended in 1982, at a conference chaired by Tommy Koh (Singapore representative), in Jamaica, and came into force in 1994 by a new round of negotiations.

The point regarding UNCLOS III is that, like every axis of geopolitical conflict, the axis of North-South geopolitical conflict[12] (triggered, in a lot, by the "Meadows Report") also took the sea as a field of dialectic (Scovazzi 2000, pp. 93–103). Therefore, at each round of discussion, "In Development" forces were weighed against "Developed" forces in the construction of a precept related to the sea and its uses in the face of a global sustainability challenge (dated from the Stockholm Declaration-1972, Principle 11, passing through Geneva-1958 and Montego Bay/1982).

Hence, the protection (present, notably, in Part XII, UNCLOS III) and exploration of the marine environment was strongly associated with the geopolitics points sensitive to developed (or fully industrialised) Nation States: the reserved areas for exploration and exploitation for developing nations, transfers technology requirements and associated with mining in common areas.

Illustratively, this preceptive dimensioning of common but differentiated responsibilities in UNCLOS III caused an impassioned speech from the United States president and was considered the pinnacle of American non-adherence to the treaty.

UNCLOS III, which was set up in 1982, meant a socio-economic breakthrough for the South but a geopolitical risk for the North. What generated its low adhesion until 1994, when it was reformed. The main provisions of this reform, in brief and summarised terms, deals with (a) the reversal of the hierarchical transit order of technical standards and decision-making power of the International Seabed Authority, with the Assembly (once sovereign) assuming a secondary position in relation to

[12] On the North-South geopolitical conflict, see Beyerlin and Marauhn (2011, pp. 8–9).

the Council; and (b) that the sharing of scientific information and the transfer of technology were no longer mandatory regarding explorations and exploitations in the sea. Only after this change has made the northern countries sign up, did UNCLOS take strength and can take its place in the pantheon of international standards of importance.

And how does that matter to SDGs, especially 14?

SDG 14 is linked, in an explicit and extraordinary way to other SDGs, to a codified international standard, that is, UNCLOS III.

At the outset, one sees a problem at the first level, namely: a soft law, reviewed in its preceptive and open format, since its format is useful to certain global needs and problems, is anchored in a definitive way and expressed, to a codified and old model of international legal norm (binding in processes, inclusive).

Deepening the syntaxes, and relying on UNCLOS III, there is a second obstacle ahead: whether the SDGs depends on the technology transfer and free transit of information, if the SDGs are strengthened as the nations are strengthened socioeconomically (Stockholm Declaration 1972), then the interpretation of methodologies 14.a and 14.c, SDG 14, disrupt both through UNCLOS.

In other words: when using the best that UNCLOS has, the dynamic now imposed points out that the transit of information obeys only the specific scope of the United Nations Educational, Scientific and Cultural Organization (UNESCO) regime (14.a), with the nations that is not being able to work with a balanced transit of technology and information: it has been relegated to private international contractual rules and to the market, under the terms of 14.c, SDG 14, cc. With the 1994 reform of UNCLOS III.

Get the idea, even though Egede (2018, pp. 182–183) did not intend to do so much, the author observed that the effect of sovereignty in the application of the ISA Mining Code rules in the Area and, with the Code, the rules about exploration and exploitation of the international seabed. In this observation, he has come up with two conclusions:

(a) The need of the private (or public, but only capitalist sub-nationals) use a sponsorship of a sovereign nation (although geopolitically it is ten or 100 times more powerful and/or influential than this nation), by the simple and logical application of the principle of sovereignty recognised and realised, under the terms of the treaties of 1648 and in the light of the duty of "No Harm" (Trail Smelter 1941)
(b) The real risk of the occurrence of "Convenience Sponsors" (in reference to the flags of convenience) that, in the light of the established in Case 17 ITLOS (on environmental responsibility of the countries sponsoring economic mining activities in the Area), risking marine biomes greatly (and SDG 14, as a consequence), just as the flags of convenience currently they do with the water depth and coastal environments (Lawand Júnior and Valdez Silva 2019, pp. 1–13). This is due to the modelling of the Sponsor State chooses by internal rules or internal administrative actions lacking due diligence and licensing duties

and betting that the peaceful conflict-resolution regime will be more favourable to non-de facto accountability.

(c) In the current format of the reform of 1994, the possibility of a few or nothing to gain, risking losing the few they have gained due to indemnities or environmental recoveries, the Sponsoring States in economically fragile stages of development, as is the case today with Tonga and Nauru.

So, like a shot that backfires a firearm, SDG 14 (and, as seen, the rest) has two weaknesses in its power language (*i.e.*, the right):

(a) If the shooting incident can be generated by forging or machining work or improper assembly—of cracks or granulation of the steel of the blocks or their incorrect fitting—, the norms (SDG and UNCLOS III) do not communicate with each other at the executive level, as they are norms that favour different principles: UNCLOS III, striving for sovereignty, and the SDGs understand the need for plurality of actors, depending on the type of problem to be faced.
(b) If the shooting incident can be generated by a defective assembly of the cartridge to be triggered, the norms (SDG and UNCLOS III) do not communicate with each other at the executive level, as they are norms that favour the transit of information and technologies in different ways: SDG, opening the traffic free of other orders' ties, UNCLOS III closing this traffic, with commercial and geopolitical variables (which, in effect, caused or prevented the advancement of solutions to the global environmental problem).

The sculpture for world peace, in the gardens of the headquarters of United Nations in New York, represented by a barrel revolver, of nineteenth century design, with a knot in the barrel, gains a new meaning in the light of SDG 14 and 14.c: an apparatus that can sabotage its own objective, the very functions, due to a structural defect: the SDGs are a new international normative type, of novel operation, aiming at the future, but which, as it is anchored in past structures and does not conceptually disengage from them, might fail.

UNCLOS III, however, well intentioned, negotiated and well done for the normative modelling that it chose to have, does not reach sufficiently perfect instruments for the solution of the terrible problem of global environmental sustainability (chosen in 72, defined in 87, and assumed in 92).

When creating the SDGs, there is a global effort to integrate tools that are capable of creating international legal content: whether by opening the international society more to new actors and its capacity in produce cooperation, prevention, science/information and governance, reading the principle of sovereignty; or because it allows the main actors in the global TO (*i.e.* the States) to integrate with the sustainability objectives more easily, by not imposing processes and the privilege of regionalisms (in terms of development and production processes).

However, when an SDG, in particular 14, supports its legal disciplines in old legal tools, it opens a window for it to prevail over the former, and does not integrate it: the mechanisms will be linked by binding standards, leading to the risk that the SDG will have to be like the principles of the Dec. of Rio-92, recognised as binding by decision of the International Court for a given subject. It also runs counter to the SDGs' raison d'être (cooperation instead of conflict).

4.3.2 *Rio/92 Declaration and Preventions, Precautions, Polluter Pays, and Comprehensive Repairs*

The judged Case 17 ITLOS integrates, obligatorily, Part XI (Area and mining) and Part XII (Protection of the marine environment), by UNCLOS, through articles 235, paragraph 2, and 22 of Annex III, to the Convention.[13]

Whether the international seabed is GPG or CHM, it is the same infrastructurally speaking. It is, therefore, important to read this decision in the light of GPGs.

The aim is the same: to safeguard the environmental values discovered or undiscovered) inherent to the Area. This safeguard can take the form of precautions and precautions or by means of repairs and compensations.

On the other hand, the damage (perspective adopted for the analysis of GPGs) concerns two legal logics: that of *nuisance* and the torts.[14] As for das nui*sance*, we observed its conceptual application in previous topics.

As for torts, the ITLOS decision (2011, pp. 31–39) abandons old concepts like the recognition of "anti-legal behaviours" and brings them closer to the concept of "behaviours capable of preventing damage".

According to ITLOS (2011, pp. 31–39), the conduct that was avoided by the offending agent but that could alert the victim, or avoid the damage, or expose the victim to the damage, will also be anti-legal. And so, they will be, regardless of the mandatory rule (general or between parties) instituted in the face of the agent. It is the interpretation of the Sponsoring State's due regard and due diligence duty which also includes the order of internal normative standards in environmental

[13] 139. Another direct obligation that gives substance to the sponsoring State's obligation to adopt laws and regulations within the framework of its legal system is set out in article 235, paragraph 2, of the Convention. This provision reads as follows: States shall ensure that recourse is available in accordance with their legal systems for prompt and adequate compensation or other relief in respect of damage caused by pollution of the marine environment by natural or juridical persons under their authority.

140. This provision applies to the sponsoring State as the State with authority over the persons that caused the damage. By requiring the sponsoring State to establish procedures, and, if necessary, substantive rules governing claims for damages before its domestic courts, this provision serves the purpose of ensuring that the sponsored contractor meets its obligation under Annex III, article 22 of the Convention to provide reparation for damages caused by wrongful acts committed during its activities in the Area.

[14] A tort is an act or omission that gives rise to injury or harm to another and amounts to a civil wrong for which courts impose liability. In the context of torts, 'injury' describes the invasion of any legal right, whereas "harm" describes a loss or detriment in fact that an individual suffers."—Legal Information Institute. Ithaca. https://www.law.cornell.edu/wex/tort. Accessed on 01 Feb. 2021.

matters as usually established by "best practices" (ITLOS 2011, pp. 41–42). This is not an ITLOS innovation, since this way of observing the occurrence of a tort was already in trend in the Cortes, as stated by Goldberg and Zipursky (2012, pp. 251–274), Calnan (2012, pp. 159–194), and Sherwin (2011, pp. 227–242).

It is fundamentally on the predictability of a conduct (provided for in an internal law, in a duty of observance, or in a duty of diligence) and its denial by the agent that the theory of torts is structured.

Calnan (2012, pp. 159–196) remarks on this sense:

> The clearest example of this encroachment — and the one most damaging to Goldberg and Zipursky's theory — appears in the element of duty. According to Goldberg and Zipursky, duty is mostly an ideal private. It establishes the terms under which accident victims are authorized to enforce against their offenders' certain norms of non injury. But for most courts, duty is not so limited. They see duty more as a public mechanism, which enables judges to create and contour negligence rules to promote the general welfare. To fulfil this function, courts faced with difficult duty questions regularly employ some form of multi-factor analysis.148 Admittedly, this analysis may include "private" considerations like the foreseeability and magnitude of the plaintiff's injury, the extent of defendant's burden of precaution.
>
> However, the analysis does not stop there. It also weighs a wide variety of public policy concerns, including, often explicitly, the duty's potential deterrent effect and the parties' respective abilities to spread the loss through insurance or otherwise.150 Thus, contrary to Goldberg and Zipursky's assertion, duty is not just about defining private wrongs, but is also about solving or ameliorating some very public problems. This social mission is no casual side job, nor are the insubstantial effects. Instead, it is a permanent and pervasive aspect of negligence's lawmaking process.

Therefore, when a mining work plan is structured in contracts with ISA, a duty of due diligence and a duty of better observance, according to an internal law of best environmental practices, informed in scientific knowledge of the exploration phase, there is, above all, a predictability of movements and procedures that necessarily bind the Sponsoring State and the sponsored entity to the duty of observance, under penalty of tort.

However, this occurrence of non-observance of conduct with the cause of damage, given the specificities of the agreed conduct, is only noticeable if the information in this plan comes to the knowledge of international actors.

In this way, we understand that the governance of a GPG, when it comes to the Area, will be much more on the information management and the practice of precaution and prevention, in binding conditions already established by ITLOS, informed by the Rio Declaration Principles-92, and assumed (*i.e.*)., in the Nauru-Tonga exploration/exploitation system.

And how to inform the content of these principles? International actors again take on their roles, coordinated by international legal norms (already seen previously in this work): science takes the lead to establish, in technical application criteria, a baseline for the recovery of the plowed area, while the international standard of the Mining Code establishes what information is access to all actors, or not, by paradigm of the Aarhus Convention. If these two circumstances are not fully established by all the actors of the mineral production process in the Area, any authorised mining is a tort, which can even be committed by ISA itself (see case ICJ—Comte Bernadotte).

Once again, the dilemma: how does a declared right to exploit and mine come up against an absolute technical impossibility resulting from another right also declared in the same qualification?

The solutions that demonstrate that this antinomy is only apparent have, throughout this work, passed through some fundamental characteristics: first, the abandonment of the only spatial perspective, for the adoption of the temporal perspective of solution of the antinomies in UNCLOS III; second, the technique (science) as a cardinal element that will nourish the hermeneutic process that will form the international legal norm.

While the other solutions already seen in this work project the mining exercise for the future, when science will determine the efficiency of an RDA in the Area, the observance of the Declaration of Rio-92 for the solution of a *tort is* linked to past temporal projections.

For this novel approach, the occurrence of qualified damage as a *tort*, and its resulting damages that have the character of restorative indemnities (Goldberg and Zipursky 2012, p. 264), should be revised according to its concept. The concept of harm is enlarged: not only actual accounting loss, but also that certain future loss, although whose calculation is uncertain or impossible to calculate, resulting from an attitude of the offending agent.

Such an attitude of the offending agent, in turn, does not only violate legal parameters, the *breach* of the offender is also marked out by technical-scientific parameters, as a way of orienting by behavioural predictability (Goldberg and Zipursky 2012, pp. 264–267).

The international actor's behavioural predictability, in the case of international seabed mining, will be exercised by other international actors—after all, no one "inhabits" the seabed, properly, but takes environmental services from it. The supervisory criterion will be the technical parameter for the assumption of operations (and, consequently, degradation) agreed upon as "acceptable".

Acceptable means that only the civil restoration of the degraded area is possible, and not indemnity due to a *tort*, as meaning similar to that of degradation.

Thus, the said damage that would characterise the *tort* in the Area, due to mining, it would be taken as the difference between what is civilly expected—and agreed as acceptable degradation, collectively suffered—and at the expense of each international actor inherent to this, and what will be contingent due to the breach of a scientific, protocol, or information rule relevant.

Therefore, in the absence of minimum predictability of reversal, *i.e.* minimum scientific knowledge of a baseline, it is impossible to exercise any management the power-duty of international dispute resolution based on existing declared responsibilities between States and sponsored mining NGOs (ITLOS 2011). How can the Area be subject to the inspection or evaluation of international actors if there are no inspection parameters? What is the size of the request made from one State to another in the face of a tort derived from a breach of conduct in international seabed mining if one does not have a basis for what has been lost from the GPGs?

It is quite illustrative that a declared right cannot be conceived without the possibility of demanding it. Thus, this fact leads us to conclude that the mining norms in the Area and the recovery of the degraded Area are not antinomic: from the perspective of responsibility in *torts*, due to a GPG, the rules are intrinsic and complementary, it does not generate its full effects without it.

Likewise, although it seems contradictory, the precautionary prevention principle is protected by a *tort* to look at the past in terms of the attitudes already taken by the potentially offending agent, in the face of an unexpected scientific criterion.

These criteria will be, necessarily, informed by international actors other than the National States and supervised with the collaboration of other actors of the same category (ITLOS 2011, passim).

Therefore, when ITLOS (2011, pp. 39–41 and 48) determines the use of the Declaration of Rio-92 as environmental criteria informing the protection standards for the marine environment contained in UNCLOS III, it is inevitable to think that the normative framework aimed at the precaution and prevention of damages in the Area depends on criteria and actors in governance process.

In a brief digression, one could look at the occurrence of the category of *tort* demonstrated not as prevention or precaution, but as polluter pays, making it difficult for the regulatory basis to support the obligations and responsibilities of the Nation State (sponsor or not) and the contracted entity.

It would not be wrong to say that a *tort and* its repair have always been traditionally linked to the polluter-pays principle (Rio Declaration 1992, Principle 15). However, the validity of the standard is not exhausted, for two reasons.

First of all, polluter pays and precaution are both principles and, as such, optimisation mechanisms in a gradual process of realisation, according to an ethical paradigm that informs the hermeneutics that will bear fruit in the norm. There is nothing that prevents points of intersection between one principle and another. In this case, governance provided this intersection: as far as I must spend to prevent damage, non-compliance with scientifically marked processes and the lack of disclosure of these processes will naturally make the obligation to protect marine environments more difficult (under the character preventive).

Second, under the risk analysis criterion, the preventive sieve can be looked at from a broader angle: several conducts that can be qualified as polluter pays will only be due to the worsening of the environmental risk of damage, increasing the value for, finally, to prevent it. It is about the syntax of one principle with the other, with prevention being an objective macro, and the polluter pays, a micro.

It remains to explore the damage itself. Goldberg and Zipursky (2012, pp. 264–267) points out that the calculation of this damage obeys would, in other times, to a theoretical criterion of the "cheapest cost avoider", *i.e.* the lowest possible value among the possible scenarios of measures to control changes in preventive management of .risks to the environment. However, we have a clear orientation from the same authors, which was more incisive and critically confirmed by Calnan (2012, p. 194), about the use of this tort in UNCLOS III: prevention and precaution, principles that do not accept cheapest as a criterion in terms of environmental value, but "the best efforts" (Sulphides Regulations, Annex 4, Section 5.1) (UNCLOS III, articles 139, 153, and 4, Annex III).

The damage from the *tort must* be recognised not as the cheapest or the most expensive, but the most appropriate set of measures to manage the new, and most serious, risk of occurrence of environmental damage in the international seabed. This is because damage and loss are different concepts, although this is contained in that: a harmful conduct may or may not generate a loss, in the sense of a loss. Keep in mind that, if a seabed mining operator does not proceed exactly as in the plan, the calculation of probabilities of the occurrence of an environmentally harmful event increases, but not necessarily that it will happen in a peremptory way. What is certain is that both generate a tort (Sherwin 2011, pp. 228–230).

Once again governance is present, guiding international society towards the optimum maintenance of a GPG: to identify this best criterion, although it is a decision of a political nature *in fine,* it is a decision that must pass through language and scientific discussion.

This circumstance is in accordance with the duty to *No Harm* and articles 31 (3) (ç) and 33—Vienna I/69: the tort exists only because there was damage, resulting from a breach of conduct on a legal or technical standard, which ultimately aims at the environmental precaution of the international seabed.

Thorough fix of the plowed area today is impossible. Scientific advances can, as far as possible, understand how to repair the Area and its ecosystems in the future. However, whether in this future of open abyssal fronts, be it now and the research phase, scientifically attested procedures and transparency are necessary in view of the environmental services provided by the Area and by the environmental heritage and the intrinsic value that it represents. The violation of these procedures generates re-calculation environmental risk and an increase in measures to prevent it.

Therefore, and in short, international seabed mining is not a contradiction in the duty to recover the mined area. In fact, and from the perspective of the responsibility of the activity, one process is intrinsic to the other: one is a requirement for the effectiveness of the other, since at the level of validity both are fully stated in precept and Court.

4.4 The Breakdown of Dialogue Between Regimes and the Emptying of Systems

The mining, by current technology, is a destructive process of the ecosystems of the geography where it is installed. This mining, for the Area, is inevitable due to the consensus between sovereignties in UNCLOS III, Part XI.

In turn, the Area consolidates characteristics of CHM and GPG, which invariably denotes:

(a) The obligation for the degrader to reconstitute the damage it causes
(b) The right of other nations to obtain from the degrader the remedial conduct of the patrimony universally declared in a binding rule
(c) In the duty of cooperation, to the extent of its powers and possibilities, of the other actors in the international community (sub-national, local, NGO, and epistemological)

The same rule that authorises mining, subject to conditions, points to a global patrimony or global public good (GPG) equally incorporated in the list of defensible rights of all National States affiliated to the UN. Dressed up as GPG, the obligation to maintain and restore the Area is a prerequisite for mining. If seen as CHM, the RDA duty is subsequent to mining.

In addition to the general Vienna I-69 rules, UNCLOS III has two specific rules of interpretation: firstly, UNCLOS III must be interpreted and practised as a single instrument (Treves 2012). In other words, the parties to the Convention do not form watertight and isolated regimes in the Law of the Sea, but necessarily syntactic with each other. Secondly, the 1994 reform is incorporated into the text of UNCLOS III but prevails over it in the event of an apparent conflict of provisions.

With these items in mind, it must be considered that Part XI (and Annex III) and Part XII must dialogue with each other to structure what will be the mining of the international seabed. Unfortunately, as seen thoroughly in this work, at the time this work is written, the points of harmonisation are minimal.

If the law is rarefied in solutions, then, even with a legal view, we must seek solutions in other fields.

We opened this work with infrastructural epistemological concepts: it was the scientific advance that allowed access to the seabed. It was the scientific advance that implied the discovery of ores in these funds, and it was the same scientific advance that provided techniques for mining them. Is it in science that the syntax between the regime for the protection of marine resources in the Area (Part XII) and the mining of these funds (Part XI and Annex III) will take place?

Mouton (1966) structured an entire reasoning not only to answer positively to the last question, but to point out that science was the most relevant direct or indirect driver of the evolution of international law in the twentieth century.

The author started from points of conflict object of international regulation to demonstrate that the balance between parties (concept on which international law is based (Aust 2010, pp. 2–8) is radically altered in the face of an epistemological quantum leap: to quote some examples of school, ecology, energy and atomic warfare, access to space beyond the earth's atmosphere, *the Internet and the WWW* they changed the equation of sovereign forces on which international law is based, to the point of changing interpretations, rules and even creating whole new regimes (Mouton 1966, pp. 269–276).

With regard to international environmental law and its inherent instrumental governance, Rei (2017, pp. 29–38) places epistemological actors in two roles: (a) reading and accessing data that attest to compliance or non-compliance with environmental standards international standards, or even the sufficiency of these norms to regulate what they propose; (b) the creation of technical-scientific parameters, *i.e.* the discovery of an imminent new ecological disaster, or of the functions of a specific population, that will motivate the evolutions in the international environmental legal norm, either by the creation of new norms through old precepts, either by creating new precepts, dealings or ways of acting.

In the case of Part XII, the order of UNCLOS III is to preserve and recover "natural resources". Although it does not use the expression "ecosystem", ITLOS, through Case 17, understands the protection of marine biomes, including that of the Area (ITLOS 2011, pp. 32–49).

Therefore, it is in the increase of the activities of the epistemological actors of international environmental law that the dialogue between the mining regimes and the environmental regimes of UNCLOS III will occur.

However, Chaps. 2 and 3, of this, bring enough information to understand that we do not have data or knowledge of the Area. Therefore, from the perspective of CHM, we will not know what to analyse or what impact we intend in order to create from the perspective of GPGs, we are not in a position to plan an environmental risk management procedure, since this risk is unknown due to what is not known, exactly, what to impact.

In an edge situation, as an illustrative way of reasoning the syntax between the Law of the Sea regimes, the option of mining without environmental care, since both are declared rights, but these obligations are not known which ones, would render the precepts of Part XII dead. for the biomes of the Area when they were finally understood (if there was even a chance).

In an opposite limit situation, the option of absolute safeguarding of the environment without mining or any other form of exploration of the Area, with the same considerations, would remove from States all the sovereign effort to agree rules to guarantee their development and occupation rights in the Area.

Either one or the other, it would render an international legal norm of an international legal system void of effect, which already suffers immense questions as to its legitimacy and effectiveness: piracy in the Horn of Africa and in Central America insular; Chinese military bases on artificial islands in the South China Sea, with attitudes incompatible with the non-appropriation of the oceans, have already placed the system of Law of the Sea represented by UNCLOS III as a contested system (Song and Tonnesson 2013, pp. 235–269).

It should be noted that this situation of non-compliance (or, at the very least, hierarchy) of the UNCLOS III norms leads to the dedication of the exact opposite of the *Whaling in the Antarctic* (ITLOS) and *Southern Bluefin Tuna* (ICJ) cases: the observance of the norms of a multilateral treaty, as a whole, and international law (fragmented, but one), derive not from the damage suffered and measurable by a State Party, but from the very sovereignty invested in it by its States Parties in the acts of negotiation, signature and ratification, which must be respected by their counterparts.

From the perspective of GPGs and international environmental law, our test balloon has more dire consequences: the right holder of the Area biome is, as explained before, humanity, if it lasts, and not the States. By act of the States, in full exercise of sovereignty, all UNCLOS III signatories transferred that ownership and undertook not to take ownership or exclude any others from the Area from access, the ISA authority being responsible for regulating this through the licensing regime, authorisations, and monitoring.

Likewise, the breach of UNCLOS III and its disrepute would take hostage the normative instruments of the SDG 2030, especially due to goal 14.c, which are important to the actions of governance and maintenance of this humanity's heritage by sub-national, locals-locals, and NGOs.

Lastly, other *commons* (CHM or GPGs) would be at risk due to the Area's previous limit, making the epistemological support of decision-making something to do without international politics, going back to concepts previous to the Treaty of Tordesillas.

Similar to it or not, it is still UNCLOS III that holds the normative bases that support the systems of international law and international environmental law and the protection of the environmental heritage in the Area. It is UNCLOS III, as an international code, that brings together 167 obedient sovereignties (according to the Vienna Convention-I-69) and subject them to its dictates.

In fact, this circumstance is solved by traditional means of law, with only a slight exchange of perspective in observing it. As explained, in Dworkin (1982), to verify a legal norm, it is necessary to observe the phenomenology of its construct, namely it is based on a precept (voted on or imposed by a constitutionally authorised issuer, i.e. a parliament). This precept is manifested in a sign that can be perceived by the applicators of law and by those presented to it (i.e. the text of the law in the Official Gazette or the King's Herald proclaiming in a public square). Then, relying on the framework of infrastructural values non-legal, said "moral" by Dworkin), the enforcer of the rule (*i.e.* a judicial court or a magistrate) hermeneutically processes this precept and produces an order (binary, "do or do not do it", "under penalty") or an optimisation mechanism towards achieving a value. Thus, there is a norm of observance in a certain collectivity or society (Dworkin 1982).

For this reason, producing a legal norm is a process that starts from present values informing a past concept and becoming something new, namely, the legal norm in force. The precept arises from an institution of power, such as a parliament, with a purpose and according to the values of this institution. This sign already has itself engraved, either in the precept itself or in the intention of a norm that it is intended to be used as a raw material, values of the time of its production. As explained, the precept can only keep concepts and express expectations that existed until then: its writing style; the vernacular used at the time; the concepts discovered, invented, or existing until then; or the intentions that prevailed at that time.

For example: In Magna Carta, when establishing the fundamentals of guaranteeing the inviolability of the home, one can read the verbal inflections of English at the time used using suffixes "th" indicating the past tense, and not "ed", as is used today any legal act from the House of Commons. Likewise, there is no mention of "person" but of "man". To indicate the holder of the right, revealing values of the time that are now considered dismal, extending this guarantee to all sexes. Finally, there is no mention of domicile, which has a concept of address, of retreating to private life

(Moraes 2017, pp. 58–60), but it uses the figure of the accession of the "house", of the construction civilian proper, in an improper metonym, indicating that, at the time, houses, manors and castles had more than a function in addition to the home and refuge of family private life (military, production, government and public functions, etc.). Nowadays, the guarantee is read not by accession but by the usefulness that a given space gives: if a container, a hotel room, or a trailer meets a household characteristic, the guarantee extends to it (Moraes 2017, pp. 58–60).

The same occurs with UNCLOS III: its precepts were shaped by a reality of conflicts of two orders, the Cold War (East-West), and the axis of conflict North-South, which opposed environmental preservation to development (1976 to 1982). The concept of sustainable development and the fall of the Berlin Wall only occurred in the late 1980s, the Section XX (87 and 89, respectively). Therefore, precepts that conceptualise or claim sustainability or the prevalence of the free flow of information are logically impossible to read in the Convention. But not to be understood as a norm defined by the treaty: note that ITLOS, in 2011, considered the Declaration of Rio-92, as explained, an instrument that integrates the treaty for mining purposes in the Area, evolving a past concept.

Our proposal, in view of the obligation to safeguard the normative precept that is sovereignly agreed (article 31 and paragraph, Vienna-I-69) and the search for meaning of the Convention as a logical system and obligatorily one (Treves 2012), is to look at this equation in time, too. But upside down.

For it to become perfect, we must see UNCLOS III as an unprecedented normative production: a precept for the future has been produced, whose infrastructural values (especially the technical-scientific) will inform it in the future and give its meaning and effectiveness.

Although Parts XI and XII are marked by styles and conceptual structures, notably from the 1970s and 1990s (in the case of the 1994 reform), they will only talk and be united if they are projected as messages for the future, when the state-of-the-art in RDA allow the drawing of baselines and techniques for the recovery of physiognomies, ecosystems, and biomes of the international seabed. Otherwise, one of the Parties will tend to cancel the effectiveness of the other and UNCLOS III will lose its unity.

Thus, we have a precept of mining and preservation of the richness of the biomes of the Area directed towards a future. This precept that, when fed with values in the time ahead, will only become effective when informed of the appropriate value. Other scientific values of RDA that could and can inform you since 1982, such as the state-of-the-art at the time of writing this work, only give the letters of UNCLOS III the efficacy of the ability to revoke previous provisions, or to produce nuisances or torts to be solved by the methods provided for in the Convention for the resolution of conflicts, for the violation of one or the other of the Parties, as in the limit example.

The equation was inverted: the precept is futuristic although done in the past. Interpretation is present.

Chapter 5
Conclusions

The present work discussed United Nations Convention on the Law of the Sea (UNCLOS III) in Part XI (mining of the international seabed) and Part XII (protection of marine environments) in the light of an obligation: environmentally recovering mined geographic regions of the Area.

The aim was to identify and understand the legal rules that allow mining in the Area and the rights and commitments of mining exploitation as one of multiple uses of the seabed: how mining exploitation can be concomitant to other uses of the seabed, considering the analysis of the obligation to recover the degraded area by mining at the mining site.

By reading UNCLOS III, supported by appropriate court judgments, which the Convention determines to be the right mechanism for resolving disputes, we observe the following in this book:

(a) The recovery of damaged area (RDA) in plowed regions of the Area is an obligation arising from the CHM nature of the Area, from its multiple uses advocated by UNCLOS III, from the prohibition of unilateral appropriation or establishment of sovereignty in the Area by the ownership of humanity, and from the characteristic of Area GPGS.

(b) RDA is a additional obligation and inherent to the mining of the Area, as it is considered a GPGS. Analysed from the perspective of the holder and the new theory of torts and nuisances, it is impossible, from the current technique, in order to establish the risk criterion and its proper management for mining in the Area aimed at the preservation of other uses and environmental heritage of use and common value.

(c) The integrating elements of UNCLOS III, notably the Rio Declaration-92 and the International Seabed Authority's (ISA's) instructions and technical studies, impose on the mining of conduct areas guided by the principles of prevention, precautionary and polluter pays, being RDA an instrument of choice. for the achievement of these principles

A. E. Lawand Junior, *Legal Aspects of the Recovery of Areas Degraded by Mining in the International Seabed*, SpringerBriefs in Law,
https://doi.org/10.1007/978-3-031-12492-1_5

(d) The RDA methodology demands a degree of knowledge of the biomes and their ecosystems that one wants to recover that one does not yet have for the Area. That is, to say, the state-of-the-art of RDA in the mined seabed is insufficient to effectively recover the mined region.

Based on findings above, observing from this RDA instrumental point of view, the research brought to light an apparent antinomy in international law, namely the compatibility of the Area and its multiple present and future uses with the mining of international seabed, both represented by normative precepts from the same legal diploma, namely Part XI and Annex III (for mining the Area) and Part XII (for the preservation of marine biomes), both from UNCLOS III.

To consider the assertions and doctrinal readings of Case 17 International Tribunal for the Law of the Sea (ITLOS) and others (arbitration, International Court of Justice (ICJ), and ITLOS) and to consider the mandatory reading one, lacking intrinsic contradictions, of the 1982 Convention, we have structured premises that a party it could not simply prevail, supplant, or sublimate the legal effects of the other.

The consequence of an option that supersedes either part of UNCLOS III unfolds:

(a) In the emptying of the regime of the sea as a whole, due to the disrepute that their dealings come to represent in the international scenario.
(b) In the emptying of other regimes that are based on CHM or GPGS concepts since their vilification would imply a few or nothing.

The solution lies in an exchange of factors: the harmonisation of the two conflicting parts of UNCLOS is possible by manipulating time, as a variable in production the formula of the legal norm: in time, sufficient knowledge of the Area to be supplied will get you RDA in mining. Until then, the mechanisms of international law and international environmental law that provide time for legal regimes must be applied.

Considering the CHM regime, moratoriums and claims for *nuisances* due to threats to the patrimony of the humanity, as well as breach of the Treaty, they would pay homage to the quota principle, guiding this regime, prevention. Governance mechanisms, in this case, would be linked to pressure groups, with the epistemological community being useful as evidence of the threat.

Considering the GPGS regime, *torts,* and normative structures for voluntary adherence (non-binding) (SDG 2030) would optimise efforts in the face of values that are also preventive, but also related to the polluter-pays principle. In this case, the governance mechanisms would take on more importance:

(a) First, they would not be just pressure groups, but would work to adopt these normative goals in their own structures, starting to function as the main actor, in their niche and power of action.
(b) Second, the epistemological community and other non-sovereigns would create scientifically informed protocols for joint and directed risk management to the Area's environment in the case of mining. The *tort would reside*, precisely, in the breach of this protocol, causing the environmental risk to oscillate, demanding

(costly) measures from all actors for the protection of—GPGS. This oscillation and consequent sacrifice would characterise the damage.

According to these instruments in the manipulation of the temporal variable of the mining process and protection of the international seabed returns in a perspective of recognizing an inversion in the post-positivist formula of the production of the legal norm: this time, a precept was not written considering the existing values at the moment, but those who are birthing at the moment when it will become fully effective in its material effects. Meanwhile, there are not enough scientific values to give coherence to the unified UNCLOS III system, through the RDA in the mined area, giving coherence and harmonising syntax of Parts XI and XII, of the Convention.

It is more honest to say, in short, that this work raises inevitable political questions to be answered, including the following.

Mankind faces a restarting an entire global chain of production processes, that is mining. However, for the first time in the history of any economic production cycle, we have the opportunity to start to adopt the best possible environmental practices to understand the ecosystem on which we will intervene before the intervention begins. Shall we take this chance?

Mining activities will inevitably start and continue, as long as we need better computers, cell phones, rockets, satellites, weapons, missiles, vehicles, buildings, and even toothpaste. Iron and copper are not enough, and the rare ores (gold, palladium, tungsten, cobalt, thorium, etc.) from the international seabed are the most abundant source of raw materials to manufacture all of this. It remains for us to know what humanity will endure: its impacts on land or its impacts on the sea? If at sea, what type and scale of impact we generate, the same ones that mining has generated on land today?

References

Abi-Saab G (1996) Cours Général de Droit International Public in Récueil de Cours, 1st edn. Martinus Nijhoff Publishers, Hague

Araújo Júnior LR (2017) A Área para além das jurisdições nacionais e seu regime jurídico. In: Borges TC et al (eds) Direito do Mar: reflexões, tendências e perspectivas, vol 1. D'Plácido, Belo Horizonte

Aust A (2010) Handbook of international law. Cambridge University Press, Cambridge

Barbosa Júnior I (2012) Oceanopolítica: conceitos fundamentais - Amazônia Azul. In: Barbosa Júnior I, More RF (eds) Amazônia Azul: política, estratégia e direito para o oceano do Brasil. Femar, Rio de Janeiro

Barbosa LM, Barbosa TC, Barbosa KC (2008) Diversificando o reflorestamento heterogêneo com espécies nativas para recuperação de matas ciliares: orientações, ferramentas e procedimentos técnico-científicos disponibilizados pelo IBt – SMA. In: Barbosa LM (ed) II Simpósio de Atualização em Recuperação de Áreas Degradadas com Ênfase em Matas Ciliares. Instituto de Botânica, São Paulo

Barboza J (1994) International liability for the injurious consequences of acts not prohibited by international law and protection of the environment. In: Collected Courses of the Hague Academy of International Law, vol 247. Hague Academy of International Law, Haia. https://doi.org/10.1163/1875-8096_pplrdc_A9789041100795_03. Accessed 10 Jan 2021

Beyerlin U, Marauhn T (2011) International environmental law. Hart Publishing, Oxford

Boyle AE (1995) Environmental regulation and economic growth. Clarendon Press, Oxford

Brasil (2016) Manual de Campanha Defesa, Química, Biológica, Radiológica e Nuclear. Ministério da Defesa - Exército Brasileiro, Brasília

Calnan A (2012) The distorted reality of civil recourse theory. Cleveland State Law Rev 60:159–194

Cançado Trindade AA (2006) International law for humankind: towards a new Jus Gentium, vol I. Martinus Nijhoff Publishers, Boston

Carlarne C (2014) Delinking international environmental law & climate change. Mich J Environ Adm Law 4(1). http://repository.law.umich.edu/mjeal/vol4/iss1/1. Accessed 09 June 2018

CLCS (Commission on The Limits of the Continental Shelf) (2019) Submissions, through the Secretary-General of the United Nations, to the Commission on the Limits of the Continental Shelf, pursuant to article 76, paragraph 8, of the United Nations Convention on the Law of the Sea of 10 December 1982. CLCS, New York. https://www.un.org/Depts/los/clcs_new/commission_submissions.htm. Accessed 29 Feb 2020

Costa de Souza HS (2019) O Direito do Mar e seus Três Nós. Arraes, Belo Horizonte

A. E. Lawand Junior, *Legal Aspects of the Recovery of Areas Degraded by Mining in the International Seabed*, SpringerBriefs in Law,
https://doi.org/10.1007/978-3-031-12492-1

Dallari Bucci M (2001) A Comissão Brundtland e o conceito de desenvolvimento sustentável no processo histórico de afirmação dos direitos humanos. In: Derani C, Costa J (eds) Direito Ambiental Internacional, 1st edn. Leopoldianum, Santos, pp 50–63

Datta S (2011) The wonderful marine world. Ministry of Information and Broadcasting of the Government of India, Nova Deli. https://www.biodiversitylibrary.org/item/252502#page/6/mode/1up. Accessed 04 Apr 2020

Denny DMT, Granziera MLM (2017) Tragedy of the commons and governance in times of emotivism. In: Rei F, Granziera MLM (eds) Global environmental issues: law and science. Leopoldianum, Santos, pp 9–27

Derani C (1997) Direito Ambiental Econômico. Max Limonad, São Paulo

Douhan A (2019) Liability for environmental damage. In: Max Planck encyclopedia of public international law. Oxford University Press, Oxford, pp 2–3. https://opil.ouplaw.com/view/10.1093/law:epil/9780199231690/law-9780199231690-e1580. Accessed 04 Apr 2020

Dulci OS (2018) Guerra fiscal, desenvolvimento desigual e relações federativas no Brasil. Revista de Sociologia e Política, No 18. Universidade Federal do Paraná, Curitiba. https://www.scielo.br/scielo.php?pid=S0104447820020001000078&script=sci_arttext. Accessed 04 Jan 2021

Dworkin R (1982) Le Positivisme. Révue Droit et Société. L.G.D.J, Paris

Egede E (2018) The area: common heritage of mankind, sponsoring states of convenience and developing states. In: Kotzur M et al (Orgs) Sustainable ocean resources governance: deep sea mining, marine energy and submarine cables. Brill Nijhoff, Leiden, pp 157–183

Evensen J (1986) Working methods and procedures in the Third United Nations Conference on the Law of the Sea. In: Collected Courses of the Hague Academy of International Law, vol 199. https://doi.org/10.1163/1875-8096_pplrdc_A9789024736225_05. Accessed 10 Jan 2021

Ferreira Filho MG (2012) Curso de Direito Constitucional. Saraiva, São Paulo

Fitzmaurice M (2002) International protection of the environment. Martinus Nijhoff, Hague

Foden J, Stuart SI, Jones AP (2009) Recovery rates of UK seabed habitats after cessation of aggregate extraction. In: Marine ecology progress series, vol 390. Inter-Research/Wiley, London, pp 15–26

Foster CE (2012) International adjudication – standard of review and burden of proof: Australia-Apples and Whaling in the Antarctic. (RECIEL) - Rev Eur Community Int Environ Law 21:80–91. Blackwell, Oxford

França TCN (2012) Nas Águas das Relações Internacionais: o oceano como meio, teatro e objetivo dos estados. In: Barbosa Junior I, More RF (eds) Amazônia Azul: política, estratégia e direito para o oceano do Brasil. Femar, Rio de Janeiro, pp 55–77

Freestone D (2018) Sustainable management of ocean ecosystems: some comments. In: Kotzur M et al (eds) Sustainable ocean resources governance: deep sea mining, marine energy and submarine cables. Brill Nijhoff, Leiden, pp 134–155

Freidah J (2019) Mining the Deep Sea. MIT - Massachusetts Institute of Technology, Cambridge. Available at: https://www.youtube.com/watch?v=MWvCtF1itQM&t=199s. Accessed 4 Nov 2020

Gautier P (2016) Le règlement des différends relatifs au droit de la mer et la protection du milieu marin. Lecture Series: Law of the Sea. UN, New York. https://legal.un.org/avl/ls/Gautier_LS.html#. Accessed 04 Apr 2020

Giddens A (2000) Mundo em Descontrole. Record, Rio de Janeiro

Giddens A (2011) Sociologia. Penso, Porto Alegre

Gikuchi L, Brouwer R, Davies J, Vidal A et al (2019) Reviving land and restoring landscapes: policy convergence between forest landscape restoration and land degradation neutrality. IUCN, Gland

Goettsche-Wanli G (2018) Sustainable production of offshore renewable energy: a global perspective. In: Kotzur M et al (eds) Sustainable ocean resources governance: deep sea mining, marine energy and submarine cables. Brill Nijhoff, Leiden, pp 8–75

Goldberg JP, Zipursky BC (2012) Rights and responsibility in the law of torts. In: Nolan DA (ed) Rights and private law, pp 251–274. https://ssrn.com/abstract=2076341. Accessed 01 Feb 2021

Gonçalves A, Costa JAF (2011) Governança Global e Regimes Internacionais. Almedina, São Paulo

Groombridge B, Jenkins MD (2002) World atlas of biodiversity. UNEP World Conservation Monitoring Centre. University of California Press, Berkeley

Grotius H (2002) The free sea. The Liberty Fund, Indianapolis

Hekinian R (2014) Sea floor exploration: scientific adventures diving into the Abyss. Springer, Londres

Hiddink JG, Jennings S, Sciberrasa M et al (2017) Global analysis of depletion and recovery of seabed biota after bottom trawling disturbance. PNAS 114(31):8301–8306. Washington: National Academy of Sciences. https://www.pnas.org/content/114/31/8301. Accessed 04 Sept 2020

Hill G, Hill K (2021) The People's Law Dictionary. Dictionary of Law. https://dictionary.law.com/Default.aspx?selected=1358. Accessed 12 Feb 2021

Honkonen T (2009) The principle of common but differentiated responsibility in post-2012 climate negotiations. Rev Eur Comp Int Environ Law – RECIEL 18(3):257–267. Oxford, Blackwell

Hossain K, Morris K (2017) Protecting arctic ocean marine biodiversity in the area beyond national jurisdiction. In: Andreone G (ed) The future of the law of the sea. Springer, Cham

ICJ (International Court of Justice) (1948) Reparation for Injuries Suffered in the Service of the United Nations in 1948. International Court of Justice, Hague. https://www.icj-cij.org/en/case/4. Accessed 01Apr 2020

ICJ (International Court of Justice) (1970) Barcelona Traction, Light and Power Company, Limited (Belgium v. Spain). Summary of the Judgement of February 5th, 1970. International Court of Justice, Hague. https://www.icj-cij.org/public/files/case-related/50/5389.pdf. Accessed 04 Jan 2021

ICJ - INTERNATIONAL COURT OF JUSTICE. Affaire des Essais Nucleaires 1974-2015. Haia: International Court of Justice, 2014. Disponível em: <https://www.icj-cij.org/public/files/case-related/59/6160.pdf>. Acesso em 2 jan.2021

ICJ (International Court of Justice) (1995) Demande d'Examen de la Situation Présentée par la Nouvelle Zélande au Titre dit Paragraphe 6;3 de l'Arrêt de la Cour de 1974 en l'Affaire des Essais Nucléaires. International Court of Justice, Hague. https://www.icj-cij.org/public/files/case-related/97/7556.pdf. Accessed 02 Jan 2021

ICJ (International Court of Justice) (2010) Whaling in the Antarctic (Australia v. Japan: New Zealand intervening). https://www.icj-cij.org/en/case/148. Accessed 31 Mar 2020

ICJ - INTERNATIONAL COURT OF JUSTICE. Whaling in the Antarctic (Australia v. Japan: New Zealand intervening), 2015. Available at: <https://www.icj-cij.org/files/case-re-lated/148/148-20140331-JUD-01-00-EN.pdf>. Access 31 Mar 2020

ISA (International Seabed Authority) (2010) Non-living resources of the Continental Shelf beyond 200 nautical miles: speculations on the implementation of article 82 of the United Nations Convention on the Law of the Sea. ISA Technical Study No. 5. Kingston

ISA (International Seabed Authority) (2012) Protection of the Seabed Environment. Kingston. https://isa.org.jm/files/files/documents/eng4.pdf. Accessed 10 Dec 2020

ISA (International Seabed Authority) (2016) Environmental Assessment and Management for Exploitation of Minerals in the Area- Isa Technical Study No. 16. International Seabed Authority, Kingston

ISA (International Seabed Authority) (2017) Report of ISA Workshop on the Design of "Impact Reference Zones" and "Preservation Reference Zones" in DeepSea Mining Contract Areas - Isa Technical Study No. 21. International Seabed Authority, Kingston

ISA (International Seabed Authority) (2019) Developing a Framework for Regional Environmental Management Plans for Polymetallic Sulphide Deposits on Mid-Ocean Ridges. - Isa Technical Study No. 21. International Seabed Authority, Kingston

ISA (International Seabed Authority) (2020) Exploration Contracts. Kingston. https://www.isa.org.jm/exploration-contracts. Accessed 20 Sept 2020

ITLOS (International Tribunal for The Law of the Sea) (1999) Southern Bluefin Tuna Cases (New Zealand v. Japan; Australia v. Japan) List of cases: No. 3 and 4. https://www.itlos.org/en/main/cases/list-of-cases/case-no-3-4/. Accessed 04 Jan 2019

ITLOS (International Tribunal for The Law of the Sea) (2011) Responsibilities and obligations of States sponsoring persons and entities with respect to activities in the Area (request for advisory opinion submitted to the seabed disputes chamber) List of cases: No. 17 Advisory Opinion of February 1st. https://www.itlos.org/fileadmin/itlos/documents/cases/case_no_17/17_adv_op_010211_en.pdf. Accessed 06 Dec 2018

ITLOS (International Tribunal for The Law of the Sea) (2015) Case 21. Advisory Opinion of 2 April 2015. https://www.itlos.org/fileadmin/itlos/documents/cases/case_no.21/advisory_opinion_published/2015_21-advop-E.pdf. Accessed 20 Jan 2020

Kaul I (2000) Biens publics globaux, un concept révolutionnaire. Le Monde diplomatique, Paris, pp 22–23. http://www.monde-diplomatique.fr/2000/06/KAUL/2359. Accessed 01 Jan 2020

Kaul I, Grunberg I, Stern M (1999) Questions and answers. In: Kaul I, Grunberg I, Stern M (eds) Global public goods: international cooperation in the 21st century. Oxford University Press, Nova York, pp 3–13

Koh T (2016) The negotiating process of UNCLOS III. https://legal.un.org/avl/ls/Koh_T_LS.html. Accessed 20 Sept 2020

Krisch N (2014) The decay of consent: international law in an age of global public goods. Am J Int Law 108(1):1–40. Cambridge University Press, Cambridge

Lawand JAE, Valdez Silva MCGP (2019) A Possibilidade Jurídica da Responsabilização dos Estados Emissores de Bandeiras de Conveniência no Direito Internacional Público por Danos Ambientais. In: Octaviano Martins EM, Oliveira PHR (eds) Direito Marítimo, Portuário e Aduaneiro: Temas Contemporâneos, vol III, 1st edn. Arraes Editores, Belo Horizonte

Lawand JAE, Valdez Silva MCGP (2020) Sobre a Observância do Princípio das Responsabilidades Comuns Mas Diferenciadas em Atividades de Mineração no Leito Marinho Internacional. In: (CBDMAR) IV Congresso Brasileiro de Direito do Mar. Centro de Estudos de Direito do Mar-USP, Santos

Legal Information Institute. Website. Cornell Law School, Ithaca. https://www.law.cornell.edu/wex/tort. Accessed 01 Feb 2021

Lodge M (2020) The legal regime for the deep seabed under Part XI of the United Nations Convention on the Law of the Sea. United Nations, Nova York. https://legal.un.org/avl/ls/Lodge_LS.html#. Accessed 01 Feb 2021

Mackinder HJ (2004) The geographical pivot of history. Geogr J 170(4):298–321. The Royal Geographical Society, Londres. https://www.jstor.org/stable/i368338. Accessed 02 Jan 2021

Marffy-Mantuano A (2011) Marine Environment Gouvernance internationale de la biodiversité marine dans une perspective de développement durable. United Nations, Nova York. https://legal.un.org/avl/ls/de-Marffy-Mantuano_LS.html#. Accessed 10 Sept 2020

Marotta Rangel V (2014) Fundos Oceânicos. In: Beirão AP et al (eds) Reflexões sobre a Convenção do Direito do Mar. Fundação Alexandre de Gusmão, Brasília, pp 463–488

Marshall Islands (2011) Deepwater horizon marine casualty investigation report. Republic of the Marshall Islands Maritime Administrator

Menezes W (2015) O Direito do Mar. Fundação Alexandre de Gusmão, Brasília

Miller KA, Thompson KF, Johnston P, Santillo D (2018) An overview of seabed mining including the current state of development, environmental impacts, and knowledge gaps. Front Mar Sci 4: 1–24. Frontiers Editorial, Lausanne. https://www.frontiersin.org/articles/10.3389/fmars.2017.00418/full. Accessed 10 Sept 2020

Moraes A (2017) Direito Constitucional. Atlas, São Paulo

Morelle RMT (2019) Deepest-ever sub dive finds plastic bag. BBC Science & Environment. https://www.bbc.com/news/science-environment-48230157. Accessed 31 May 2020

Mouton MW (1966) The impact of science on international law. In: Collected Courses of the Hague Academy of International Law, vol 119. Martinus Nijhoff, Haia

Myrdal A (1971) Preserving the oceans for peaceful purposes. Martinus Nijhoff, Haia

Nye JS Jr (2002) The American National Interest and the global public goods. Int Aff 78(2): 233–244. Oxford University Press, Oxford

Octaviano Martins EM (2016) Segurança Marítima e Desenvolvimento Sustentável. In: Gonçalves A, Rodrigues GMA (eds) Direito do Petróleo e Gás: marco regulatório, aspectos ambientais e internacionais. Leopoldianum, Santos, pp 211–226

Oda S (1969) International law of the resources of the sea. In: Collected Courses of the Hague Academy of International Law, vol 127. Martinus Nijhoff, Haia

OECD (Organisation for Economic Co-operation and Development) (2016) Workshop on maritime clusters and global challenges. OECD, Paris

Ostrom E (1990) Governing the Commons: the evolution of institutions for collective action. Cambridge University Press, Cambridge

Piketty T (2013) Le Capital au XXIe Siècle. Seuil, Paris

Rei F (2017) International environmental law: approaches concerning the influence of science and technology. In: Rei F, Granziera MLM (eds) Global environmental issues: law and science. Leopoldianum, Santos, pp 29–38

Reis RGG, Almeida FEA (2012) Luta pelo domínio do mar na História. In: Barbosa Junior I, More RF (eds) Amazônia Azul: política, estratégia e direito para o oceano do Brasil. Femar, Rio de Janeiro, pp 23–54

Rittel HWJ, Webber MM (1973) Dilemmas in general theory of planning. Policy Sciences, No. 4. Elsevier, Amsterdam, pp 155–169

Santos MM, Jorge PAS, Coimbra J et al (2018) The last frontier: coupling technological developments with scientific challenges to improve hazard assessment of deep-sea mining. Sci Total Environ 62:1505–1514. Elsevier, Amsterdam

Schrijver N (2008) The evolution of sustainable development in international law: inception, meaning and status. Martinus Nijhoff Publishers, Leiden

Scovazzi T (2000) The evolution of international law of the sea: new issues, new challenges. In: Collected Courses of the Hague Academy of International Law. Martinus Nijhoff Publishers, Leiden

Sherwin E (2011) Interpreting tort law. Cornell Law Faculty Publications, article 813. Ithaca: Cornell Law Faculty Publications, paper 813, pp 227–242. http://scholarship.law.cornell.edu/facpub/813. Accessed 01 Feb 2021

Singer P (1975) Curso de Introdução à Economia Política. Forense, Rio de Janeiro

Singh N (1973) International law: problems of merchant shipping. Stevens & Sons, Londres

Slaughter AM (1998) International law and international relations. Martinus Nijhoff Publisher, Haia

Song Y, Tonnesson S (2013) The impact of the law of the sea convention on conflict and conflict management in the South China Sea. Ocean Dev Int Law 44(3):235–269. Taylor and Francis, London

Stern B (2000) La Sucession d'États. Recueil des Cours. T. 262. Martinus Nijhoff Publisher, Haye

Stiglitz JE (1999) Knowledge as a global public good. In: Kaul I, Grunberg I, Stern MA (eds) Global public goods: international cooperation in 21st century. Oxford University Press, Nova York, pp 308–325

Tanaka Y (2018) Toward sustainable management of marine natural resources. In: Kotzur M et al (eds) Sustainable ocean resources governance: deep sea mining, marine energy and submarine cables. Brill Nijhoff, Leiden, pp 110–133

Tavares SRL (2008) Áreas Degradadas: Conceitos e Caracterização do Problema. In: Tavares SRL (ed) Curso de Recuperação de Áreas Degradadas - A Visão da Ciência do Solo no Contexto do Diagnóstico, Manejo, Indicadores de Monitoramento e Estratégias de Recuperação. Embrapa Solos, Rio de Janeiro, pp 1–8

Taylor ML, Roterman C (2017) Invertebrate population genetics across Earth's largest habitat: the deep-sea floor. Mol Ecol 26(19):4872–4896. Wiley, Hoboken. https://onlinelibrary.wi-ley.com/doi/10.1111/mec.14237. Accessed 07 Sept 2018

Tharpes YL (1989) International environmental law: turning the tide on marine pollution. Inter-Am Law Rev 20(3):579–614. University of Miami, Miami

Treves T (2012) The law of the sea on the thirtieth anniversary of the United Nations Law of the Sea Convention. United Nations, Nova York. https://legal.un.org/avl/ls/Treves_LS.html. Accessed 12 Dec 2021

Truyol y Serra A (1965) L'Expansion de la Société Internationale aux XIXe et XXe Siècles. Recueil de Cours de l'Académie de Droit International de La Haye. Martinus Nijhoff Publishers, Haye, pp 87–179

UN (United Nations) (1987) Report of the World Commission on Environment and Development: our common future. UN, Nova York. https://sustainabledevelopment.un.org/content/documents/5987our-common-future.pdf. Accessed 12 Dec 2020

Van Dover CL et al (2016) Application of biological studies to governance and management of the deep sea. In: Clark MR et al (eds) Biological sampling in the deep sea. Wiley-Blackwell, Hoboken, pp 404–430

Vonnahme TR, Molari M, Janssen F et al (2020) Effects of a deep-sea mining experiment on seafloor microbial communities and functions after 26 years. Sci Adv 6(18):1–14. AAAS, Pensilvânia. https://www.science.org/doi/epdf/10.1126/sciadv.aaz5922. Accessed 4 Nov 2020

Wedding LM, Reiter SM, Smith CR et al (2015) Managing mining of the deep seabed. Sci Mag 349(6244):144–145. https://doi.org/10.1126/science.aac6647. AAAS, Washington

Weiss EB (1992) Intergenerational equity: a legal framework for global environmental change. In: Weiss EB (ed) Environmental change and international law: new challenges and new dimensions. United Nations University Press, Tokyo, pp 385–412

Wolfrum R (1999) Means of ensuring compliance with and enforcement of international environmental law. Martinus Nijhoff, Haia

Wolfrum R (2012) Common heritage of mankind. In: Max Planck encyclopedia of public international law. Max Planck Institute for Comparative Public Law and International Law, Heidelberg. http://www.mpepil.com. Accessed 02 Apr 2020, pp 1–8

Wolfrum R (2018) Development under the law of the sea convention. In: Kotzur M et al (eds) Sustainable ocean resources governance: deep sea mining, marine energy and submarine cables. Brill Nijhoff, Leiden, pp 101–109

Zou K (2018) State practice in deep seabed mining: the case of the People's Republic of China. In: Kotzur M et al (eds) Sustainable ocean resources governance: deep sea mining, marine energy and submarine cables. Brill Nijhoff, Leiden, pp 185–203

The manufacturer's authorised representative in the EU is Springer Nature Customer Service Centre GmbH, Europaplatz 3, 69115 Heidelberg, Germany. If you have any concerns regarding our products, please contact ProductSafety@springernature.com

Printed and bound by CPI Group (UK) Ltd, Croydon, CR0 4YY
08/10/2026
02230056-0004